The Universe, the Electron, and the Vacuum

A Universal Scale Factor that Covers All the Universe's Lifetime
Directly Links the Big Bang with the Present Universe
A Big Dip in the Hubble Constant
A New Interaction Guiding Universe Expansion
Universe Expansion as a Type of Vacuum Polarization
The Universe as a Particle

Stephen Blaha Ph. D.
Blaha Research

Pingree-Hill Publishing
MMXIX

Rev. 00/00/01   June 24, 2019

To My Parents

Some Other Books by Stephen Blaha

All the Megaverse! Starships Exploring the Endless Universes of the Cosmos using the Baryonic Force (Blaha Research, Auburn, NH, 2014)

SuperCivilizations: Civilizations as Superorganisms (McMann-Fisher Publishing, Auburn, NH, 2010)

All the Universe! Faster Than Light Tachyon Quark Starships & Particle Accelerators with the LHC as a Prototype Starship Drive Scientific Edition (Pingree-Hill Publishing, Auburn, NH, 2011).

Cosmos Creation: The Unified SuperStandard Model, Volume 2, SECOND EDITION (Pingree Hill Publishing, Auburn, NH, 2018).

Immortal Eye: God Theory: Second Edition (Pingree Hill Publishing, Auburn, NH, 2018).

Unification of God Theory and Unified SuperStandard Model THIRD EDITION (Pingree Hill Publishing, Auburn, NH, 2018).

Calculation of: QED α = 1/137, and Other Coupling Constants of the Unified SuperStandard Theory (Pingree Hill Publishing, Auburn, NH, 2019).

Coupling Constants of the Unified SuperStandard Theory SECOND EDITION: We Find the Fine Structure Constant 1/137.0359801, and so: OUR UNIVERSE AND LIFE! Also a Universal Eigenvalue Function for all Known Interactions, And Running Coupling Constants to all Perturbative Orders (Pingree Hill Publishing, Auburn, NH, 2019).

Available on Amazon.com, bn.com Amazon.co.uk and other international web sites as well as at better bookstores (through Ingram Distributors).

CONTENTS

FIGURES and TABLES

INTRODUCTION

The recent flurry of reports suggesting difficulties with our understanding of the evolution of the universe has led the author to consider a new universal scale factor that specifies the expansion of the universe. Remarkably, the universal scale factor well describes the universe's growth in the large starting from the Big Bang up to the current time. But, surprisingly, the growth is punctuated by a major dip, the Big Dip, in universe size about the time of the transition from a radiation-dominated to a matter-dominated universe.

The implications of the universal scale factor are presented in a large number of plots of quantities such as the Hubble Constant, the scale factor, universe density and universe pressure as well as related quantities.

After describing the universal scale factor theory the author describes the justification for its form as an analogue of electron vacuum polarization—if one views the entire growth history of the universe as a whole.

Based on this analogy the book suggests the existence of a new force (in some detail) that "guides" the growth of the density of the universe, and thereby determines its expansion.

1.The State of the Universe

The universe is vast and mysterious. New phenomena are being reported almost daily. Recently there has been a surge of interest in the growth of the universe due to the apparently increasing value[1] of the Hubble Constant H(t). The cause of the increasing expansion rate is not known. This book, and Blaha (2019c), suggests that the standard approach to calculating the scale factor of the universe and the Hubble Constant is not correct because it neglects the large pressure that persists in the universe since the earliest times.

We will suggest a new universal form for the scale factor that leads to a satisfactory picture of the universe in the large and accounts for its expansion. Based on it, we will calculate the time dependence of the density and pressure in the universe. We will also show the Hubble Constant has the currently known values and extends backward in time to the end of a Big Bang metastate at 10^{-165} second..

We will propose a fundamental justification for the universal scale factor based on an analogy with vacuum polarization and the Renormalization group. The justification of the new universal scale factor suggests that the growth of the universe is determined by the time dependent energy density of the universe which is in turn determined by a new unknown interaction. (The evolution of the universe is usually viewed as the result of General Relativity, energy conservation and the equation of state.) The new interaction "manages" the expansion of the universe where the expansion of the universe can be viewed as a form of "ongoing" vacuum polarization analogous to the vacuum polarization of an electron.

Based on this view we view the universe as a type of particle (not elementary) in a larger Megaverse of many dimensions.

---

[1] A. Riess *et al*, The Astrophysical Journal **875**, 145 (2019) and references therein.

The universal scale factor form leads to a Big Dip of the universe in which the universe contracts for a time. The scale factor and the Hubble Constant correspondingly dip.

Thus the universe does not appear to be continuously expanding but has a period when it contracts. Interestingly the dips appear near the time of transition from a radiation-dominated universe to a matter-dominated universe. One might view the universe transition as a change to a more compact form.

2. The Problem of Determining the Evolution of the Universe

The Robertson-Walker Model implements a form of Einstein's Theory of General Relativity. At first glance it appears to be a complete theory. However it is not complete because it does not specify the energy-momentum density $\rho(t)$ or the pressure $p(t)$ although it does assume an isotropic, homogeneous universe.

The Robertson-Walker metric is

$$d\tau^2 = dt^2 - a^2(t)k^{-1}[dr^2/(1 - kr^2) + r^2(d\theta^2 + \sin^2\theta \, d\varphi^2)] \qquad (2.1)$$

2.1 Scale Factor Form based on the Energy Conservation Equation

The energy conservation equation expressed in terms of the scale factor $a(t)$ in the Robertson-Walker Model is

$$d(\rho a^3)/da = -3pa^2 \qquad (2.2)$$

Typically the evolution of the universe is viewed as consisting of four phases: the Big Bang, the radiation-dominated phase, the matter-dominated phase, and the exponential "explosive" phase with each of the last three phases corresponding to a term in the standard form for the total energy density ρ_{tot}:

$$\rho_{tot}(t) = \rho_{crit}[\Omega_\gamma(t) + \Omega_m(t) + \Omega_\Lambda(t)] \qquad (2.3)$$

where[2]

$$\rho_{crit} = \text{Critical density} = 1.87840(9) \, h^2 \times 10^{-29} \, \text{g/cm}^{-3}$$

---

[2] M. Tanabashi *et al* (Particle Data Group), Phys. Rev. D**98**, 030001 (2018).

Ω_Λ = Dark Energy density/ρ_{cr} = ρ_{de}/ρ_{cr} = 0.692 ± 0.012

Ω_d = Cold Dark matter density/ρ_{cr} = ρ_c/ρ_{cr} = 0.1186(20) h^{-2}

Ω_b = Baryon density/ρ_{cr} = ρ_b/ρ_{cr} = 0.02226 h^{-2}

$\Omega_m = \Omega_b + \Omega_d$

= pressureless Matter density/ρ_{cr} = ρ_m/ρ_{cr} =0.308 ± 0.012

Ω_γ = radiation density/ρ_{cr} = ρ_γ/ρ_{cr} = 2.473h^{-2} × 10^{-5} (T/2.7255)$^4 h^{-2}$

= 5.38 × 10^{-5}

and

h = the Hubble parameter = 0.678(9)

In the radiation-dominated phase when the pressure is negligible eq. 2.2 implies

$$a(t) \propto t^{1/2} \qquad (2.4)$$

In the matter-dominated phase eq. 2.2 implies

$$a(t) \propto t^{2/3} \qquad (2.5)$$

In the Ω_Λ-dominated (explosive) phase eq. 2.2 implies

$$a(t) \text{ is an exponential in t} \qquad (2.6)$$

Blaha (2019c) gives the expressions for a(t0 which are summarized in Figs. 2.1 and 2.2.

Figs. 2.1and 2.2 also shows the Big Bang Metastate phase that was found based on generalizing the Robertson-Walker Model to a quantum theory[3] in Blaha (2004) and Blaha (2019c).

---

[3] The concept behind the quantum generalization was based on an analogy to the hydrogen atom where quantum theory stabilizes electron orbits preventing the collapse to the nucleus found in the classical hydrogen model. Without quantization the Big Bang is also strewn with infinities.

Epoch	Type	Phases	Time Period
I	Explosive Growth	Dark Energy-dominated	$2.87 \times 10^{17}\,\text{s} - 4.35 \times 10^{17}\,\text{s}$
II	Expanding	Matter-dominated	$7 \times 10^{11}\,\text{s} - 2.87 \times 10^{17}\,\text{s}$
		Radiation-dominated	$1.26 \times 10^{-165}\,\text{s} - 7 \times 10^{11}\,\text{s}$
III	Metastable Big Bang	Blackbody Y quanta dominated	$0\,\text{s} - 1.26 \times 10^{-165}\,\text{s}$

Epoch	Type	Phases	Time Interval
I	Explosive Growth	Dark Energy-dominated	4.7 Gyr
II	Expanding	Matter-dominated	9.1 Gyr
		Radiation-dominated	2.22×10^{-5} Gyr
III	Metastable Big Bang	Blackbody Y quanta dominated	1.26×10^{-165} sec

Figure 2.1. Phases of the Universe.

t	Time	Re a(t)	Im a(t)	a(t, ř)
t_{now}	13.8 Gyr	1	-6.38×10^{-93}	
Explosive				$\exp(H_0\Omega_\Lambda^{\frac{1}{2}}t)$
t_E	9.1 Gyr	,76		
Matter				$2H_0\Omega_m^{\frac{1}{2}}t)^{2/3}$
t_{RM}	2.2×10^{-5} Gyr	1.8×10^{-93}		
Radiation				$(2H_0\Omega_\gamma^{\frac{1}{2}}t)^{\frac{1}{2}}$
T_c	1.26×10^{-165} s	6.38×10^{-93}		
Big Bang metastate				Re $a_{BBRW}(t, ř) = (\gamma + a(t))/2$ Im $a_{BBRW}(t, ř) = (\gamma + a(t))\,ř/2$
0	0 s	3.19×10^{-93}	-3.19×10^{-93}	
?				

Figure 2.2 Times and scale factors of the phases after Blaha (2019c)..

2.2 The Quantum Big Bang Metastate

We developed a detailed theory of the quantum Big Bang metastate in Blaha (2004) and (2019c). This theory has no infinities and gives an ultra-short lived metastate. Fig. 2.3 summarizes its features.

2.3 The Problem of the Standard Approach to Calculating the Expansion of the Universe

Although we found what appears to be a satisfactory initial Big Bang metastate, problems arose when we attempted to calculate the evolution of the universe after the metastate. The problems are best illustrated by the plot of the Hubble Constant H(t) in Fig. 2.4 for various parameter values. The Hubble Constant squared was negative between years 380,000 to the present.

After examining our calculation we believe the source of the problem is in the use of the matter-dominated scale factor in the calculation—even though the time interval was mostly in the matter-dominated phase.

We concluded that the calculation of scale factors based on the energy equation eq. 2.1 is not satisfactory. As a result we suggested a universal form for the scale factor that leads to satisfactory results for the entire history of the universe.[4]

Time	0	t_c
Phase	**Big Bang metastate Beginning**	**Big Bang Metastate End**
Time	0	1.26×10^{-165}
Re a(t)	3.19×10^{-93}	8.166×10^{-93}
Im a(t)	-3.16×10^{-93}	-5.24×10^{-93}
Re radius	4.278×10^{-65} cm	8.5×10^{-65} cm
Im radius	4.278×10^{-65} cm	8.5×10^{-65} cm
Volume	4.37×10^{-192} cm^3	2.6398×10^{-191} cm^3
Central Expansion Energy Density	1.63×10^{218} GeV/cm^3	4.08×10^{217} GeV/cm^3
Edge Expansion Energy Density	8.16×10^{217} GeV/cm^3	2.04×10^{217} GeV/cm^3
Total Expansion Energy[5]	5.34×10^{35} eV	8.07×10^{35} eV
Hubble Constant[6] **(km s^{-1} Mpc^{-1})**	1.79×10^{218}	1.14×10^{126}

Figure 2.3 The Big Bang Metastate detailed data.

---

[4] This form of the scale was proposed in Blaha (2019c).

[5] The expansion energy does not include the mass-energy of particles in the Big Bang universe. It only includes Y field black body energy which drives the initial expansion. All particles are massless initially and all interparticle forces are zero.

[6] In the Big Bang metastable state we display the Hubble constant at the "expanding' edge.

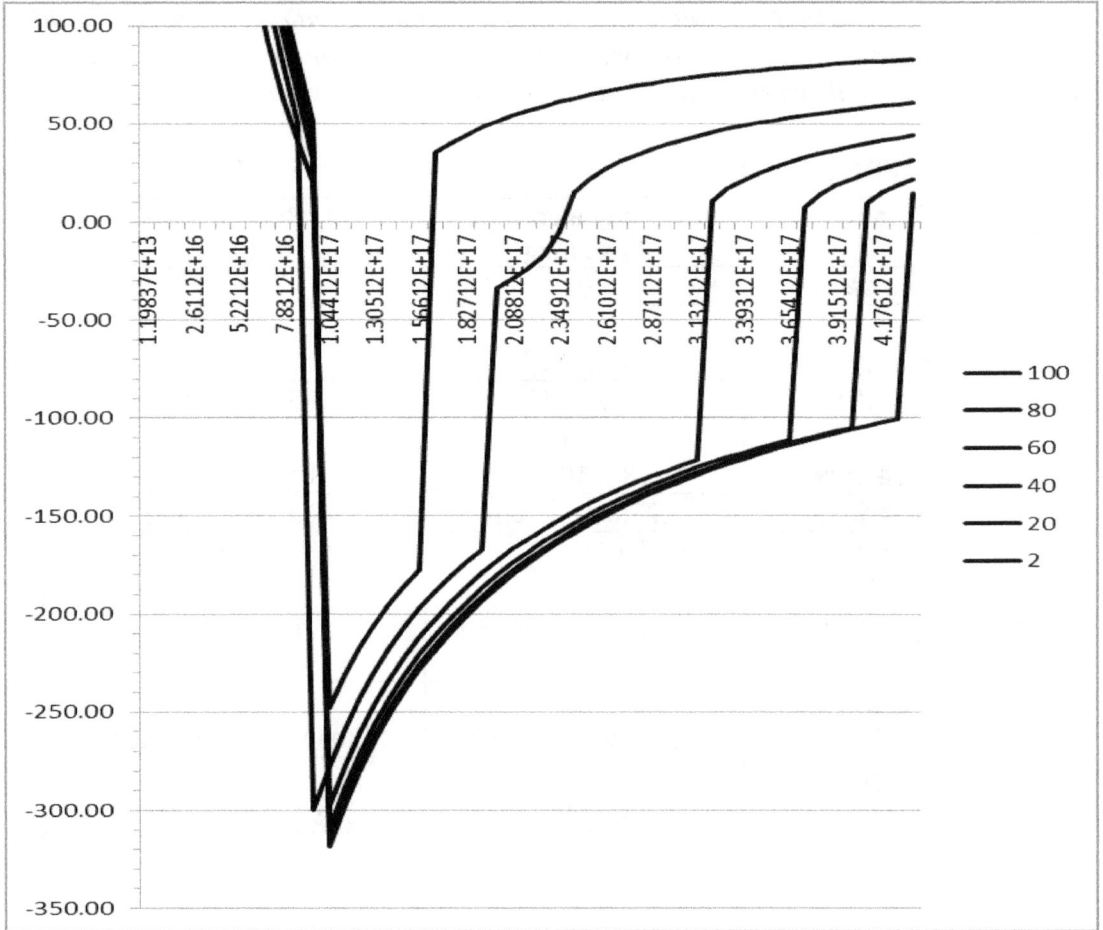

Figure 2.4. Plot of H(t) as a function of time in seconds for six choices of a parameter. The apparently negative values of H(t) signal the fact that H(t) squared is negative at those points due to the dominance of the $c^2k/a^2(t)$ term in the Einstein equation.

3. A Universal Scale Factor for the Evolution of the Universe

In this chapter we define a universal scale factor that fits the known values of the Hubble Constant, yields a value for the Hubble Constant consistent with our Big Bang metastate Hubble Constant values, and has a form that makes physical sense.

The Robertson-Walker metric is

$$d\tau^2 = dt^2 - a^2(t)k^{-1}[dr^2/(1 - kr^2) + r^2(d\theta^2 + \sin^2\theta\, d\varphi^2)] \qquad (3.0)$$

a(t) is a function of time called the scale factor. Since a(t) is dimensionless there must be at least a second parameter with the dimension of time. We will call the parameter T. Thus the scale factor a(t) has the parameters t and T:

$$a(t, T)$$

although we will simply express it as a(t) generally.

This chapter calculates the growth of the universe based on a phenomenological approximation[7] to the growth of the universe from the Big Bang metastate to the present. It addresses the current impasse in our understanding of universe expansion.

Clearly something is wrong with our recent understanding of the universe. Experimental data tells us this.[8] A further suggestion of this problem appears in the negative values of the Hubble Constant squared in our calculations in chapter 20 of Blaha (2019c) which appears to be more than a result of approximations.[9]

---

[7] See Blaha (2019c).

[8] See A. Riess *et al*, The Astrophysical Journal **875**, 145 (2019) and references therein.

[9] The primary cause of the problem may be the use of the matter-dominated a(t) as the initial input for the calculation of the true a(t) taking account of a Megavers surface tension input term. However there appears to be more issues that lead to the problems with the Hubble Constant calculation.

In this chapter we begin with a look at the Hubble Constant expansion experimental data. We then proceed to suggest a phenomenology that accounts for the data and nicely projects back to the Big Bang metastate discussed earlier in Blaha (2019c) (and in Blaha (2004)). A justification for this phenomenological fit is provided in chapter 5.

3.1 Hubble Universe Expansion Experimental Data

There are a number of astrophysical studies of the universe that suggest that the Hubble Constant is *not* constant. Although there are significant margins of error it appears that the early universe "beginning" epoch around 380,000 years had a Hubble Constant of 67.8 km s^{-1} Mpc^{-1}.[10] More recently, red shift studies of quasars have given a Hubble Constant of 73.2 km s^{-1} Mpc^{-1}.[11] And studies of binary black hole merger gravity waves have given a Hubble Constant of 75.2 km s^{-1} Mpc^{-1} (and earlier of 78 km s^{-1} Mpc^{-1}). Another study of events at 1.8 billion ly yielded a Hubble Constant of 70.0 km s^{-1} Mpc^{-1}.[12] Further studies have given the Hubble Constants: 1. Of variable stars 73.2 km s^{-1} Mpc^{-1}, 2. Of light bent by distant galaxies 72.5 km s^{-1} Mpc^{-1}, and 3. Of Magellan Cepheids 74.03 ± 1.42 km s^{-1} Mpc^{-1}. [13]

The only apparent conclusion at this time is that there was a Hubble Constant (Constant) H of approximately 67.8 km s^{-1} Mpc^{-1} early in the universe, and of perhaps 75.2 km s^{-1} Mpc^{-1} at the current time. Thus an increasing Hubble Constant.

For the purpose of discussing the apparent increase in H with time, we average the above seven "recent" values of H in the spirit of Bayesian equal probability to obtain a recent time Hubble average of 73.7. Thus there appears to be a 9% increase in the Hubble Constant over time.

---

[10] See, for example, K. Aylor *et al*, arXiv:1811.00537v1 (2018) based on studies of the cosmological sound horizon.
[11] M. Soares-Santos *et al* , arXiv:1901.01540 (2019).
[12] B.P. Abbott *et al*, arXiv:1710.05835 (2017).
[13] J. T. Nielsen *et al*, Marginal evidence for cosmic acceleration from Type Ia supernovae, Nature Scientific Reports (2016); arXiv:1506.01354 (2015)

3.2 A Phenomenological Fit to the Known Hubble Constant Data

The radiation-dominated and the matter-dominated scale factors a(t) both are power laws in time as seen earlier in Blaha (2019c). We therefore will assume that the true a(t) has a power law form:

$$a(t) = (t/t_{now})^{g + ht} \tag{3.1}$$

where g and h are constants. (The constant h is *not* the Hubble parameter.) There is an "ht" term in the exponent based on the rise in H(t) noted above in the experimental data. The Hubble Constant implied by eq. 3.1 is

$$H(t) = (da/dt)/a = g/t + h(1 + \ln(t/t_{now})) \tag{3.2}$$

If we set the value of H(t) at two values of time, then g and h are determined. Based on the above discussion we use the experimental data:

$$H(t_c) \equiv H(380,000 \text{ yr}) = 67.8 \tag{3.3}$$
$$H(t_{now}) = 73.7$$

Eqs. 3.2 and 3.3 imply

$$h = (t_c H(t_c) - t_{now}H(t_{now}))[\; t_c - t_{now} + t_c \ln((t_c/t_{now})]^{-1} \tag{3.5}$$
$$g = (H(t_{now}) - h) \; t_{now}$$

Substituting the parameter values of eq. 3.3 we obtain

$$h = 2.27402 \times 10^{-18} \tag{3.6}$$
$$g = 0.000283993 = 2.83993 \times 10^{-4}$$

Since

$$ht = .9896 \; t/t_{now} \cong t/t_{now}$$

an alternate possible form for a(t) is

$$a(t) = (t/t_{now})^{g + t/t_{now}} \tag{3.1a}$$

We will use eqs. 3.1 and 3.5 for calculations and in plots unless explicitly stated otherwise.

We use these parameters to calculate $a(t)$, $H(t)$, and Ω_T in the following plots at the end of the chapter for the time ranges: 1) from 380,000 years to the present, and 2) from the Big Bang metastate at $t = 1.1984 \times 10^{-165}$ s to the present.

The times are displayed in seconds and gigayears (Gyr). In some plots a logarithmic plot was necessary because of the vast ranges of values. *We now turn to comments on the plots since some new and unforeseen features emerge.*

The quantity Ω_T is the excess of energy and pressure beyond that specified by Ω_γ, Ω_M, and Ω_Λ in the conventional expressions for the matter-energy density. The calculation of Ω_T which is in part energy density and in part pressure,(and possibly in part a Megaverse energy influx), is based on the equation

$$\Omega_T = (H(t)^2 + c^2k/a^2)/H_0{}^2 - \Omega_\gamma/a^4 - \Omega_M/a^3 - \Omega_\Lambda \tag{3.7}$$

using the Einstein equation

$$H(t) = (da/dt)/a(t) = [H_0{}^2\rho_{tot}(t)/\rho_{cri} - c^2k/a^2(t)]^{\frac{1}{2}}$$

Other quantities of interest are

$$d\Omega_T/dt = H(t)[-2g/(H_0{}^2t^2) + 2h/(H_0{}^2t) - 2c^2k/a^2 + 4\Omega_\gamma/a^4 + 3\Omega_M/a^3] \tag{3.8}$$

$$d(\rho R^3)/dR = -3pR^2 \tag{3.9}$$

$$p = -(da/dt)^2\rho - a\, da/dt\, d\rho/dt \tag{3.10}$$

$$p = -a(t)^2(H(t)^2 + 1/3\, H(t)\, d\rho/dt) \tag{3.11}$$

where c is the speed of light, $k = 5.56 \times 10^{-57} \, cm^{-2}$, ρ is the total energy-matter density, p is the pressure, Ω_T is the excess energy density, and

$$R = k^{-\frac{1}{2}} \, a(t) \tag{3.12}$$

3.3 Smooth H(t) Connection to Big Bang Metastate

The parameters in eqs. 3.1 and 3.2 were set by the H(t) data (eq. 3.5) at 380,000 years and the present. If we extrapolate back to the end of the Big Bang metastate then we find a good match between the values of a(t) and H(t) of the Big Bang metastate and the extrapolation:

		H(t)
Big Bang Metastate	Big Bang Center	$8.95 \times 10^{217} \, km \, s^{-1} \, Mpc^{-1}$
	Big Bang Edge	$1.14 \times 10^{126} \, km \, s^{-1} \, Mpc^{-1}$
Extrapolated H(t)		$7.678 \times 10^{180} \, km \, s^{-1} \, Mpc^{-1}$

where the Big Bang values appear in section 13.6.2 of Blaha (2019c).. Note that the extrapolated value is within the range of values in the Big Bang metastate and is roughly an average of the range of Big Bang values. Thus our H(t) fit (eqs. 3.1 and 3.2) extends smoothly back to the Big Bang.

The scale factor also shows a rough agreement. Using

$$\text{Re } a_{BBRW}(t) \cong \gamma/2 + a(t)/2 \tag{13.3.2.17a}$$

from Blaha (2019c) where

$$\gamma/2 = 8.16 \times 10^{-93}$$

we see that

$$\text{Re } a_{BBRW}(t) \cong a(t)/2$$

giving us agreement in a(t) within a factor of two.

Thus H(t) and a(t) are in a rather remarkable approximate agreement for the Big Bang metastate and our extrapolation at the cross over time to the radiation-dominated phase 1.1984×10^{-165}s.

3.4 Fluctuating Behavior of a(t) and H(t) – The Big Dip

An examination of Figs. 3.1, 3.2 and 3.3 reveal a Big Dip in H(t) which seems to have been unforeseen in astrophysical investigations. The Big Dip has a corresponding big peak in $\Omega_T(t)$ as shown in Fig. 3.5. The big peak occurs at approximately the same point in time as the Big Dip. Fig. 3.4 shows a corresponding minimum for a(t).

The reason for these anomalies is somewhat uncertain although it is apparent that they are not a fault of the scale factor fit in eq. 3.1. A similar problem (negative H^2) appears in the calculations of chapters 17 through 20 of the conventional calculation in Blaha (2019c).

3.4.1 Location of the Big Dip in H(t)

The locations in time of the Big Dip events are:

Big Dip low point at $t = 8.71 \times 10^{13}$ s.
Ω_T plot peak at $t = 8.71 \times 10^{13}$ s

Since the times match it appears that the Big Dip is a result of massive growth in the mass-energy (and pressure) embodied in Ω_T.

Since the changeover from a radiation-dominated phase to a matter-dominated phase occurs at a slightly earlier time:

Radiation – Matter Domination Transition: $t = 1.48 \times 10^{12}$ s.

it seems reasonable to conclude the transition from radiation-dominated to matter-dominated causes the Big Dip to occur. The matter-dominated phase transition causes shrinkage as shown in a(t) in Fig. 3.4. The universe contracts! [14] The time delay between

---

[14] Rather like the condensation of water vapor to liquid.

the transition and the Big Dip may be attributed to the time required for the transition to occur. (The universe is large at this time after all)

3.4.2 Overshoot in H(t)

The result of the Radiation-Matter transition is negative H(t) for the energy density Ω_T. H(t) "overshoots" and becomes negative. Crudely put, the clumping of matter in the matter-dominated phase appears to introduce a compactness that results in a decrease in universe size.

3.5 Mystery of the Big Dip in H(t) - A Scenario

At the Big Dip H(t) changes from a declining to a rising trajectory. Based on this fact and the Big_Bang-Megaverse_Driven model presented in Blaha (2019c) the following scenario seems reasonable:

1. The initial peak, and immediate decline, in H(t) is due to the Y black body radiation phase pressure that decreases rapidly after the Big Bang metastate ends. Thus Ω_T declines rapidly (Fig. 3.7) with the Y pressure decline. (Note Ω_T is a sum of energy density and pressure.)

2. The peak in Ω_T in Fig. 3.5 reflects an influx of energy (from the Megaverse?) that causes H(t) to begin increasing. There is also a dip below zero in H(t) in Fig.3.3 signifying the shrinkage of the universe as a(t) in Fig. 3.4 shows.

3. Afterwards Ω_T continues to be significant and increasing as a(t) and H(t) rise to the present time.

4. In the future Ω_T should continue to rise. The energy increase that this situation implies suggests a certain reality to the part of our Big_Bang-Megaverse_Driven Model where the surface tension force causes a significant influx of mass-energy from the Megaverse.

3.6 Current H(t() Increase Explained

The scenario presented in the previous section "explains" the rise of H(t) from 67.8 to an average 73.7.

3.7 An Interlude in the Eons

Based on the above analysis it appears that the universe is currently in an *interlude* following a decline in growth rate after the Big Bang, and the new beginning of major growth due to Megaverse energy influx that will last for eons.

3.8 Scaling in the Scalar Scale Factor: Changing the Base Time from t_{now} to a different Base Time

The use of $T = t_{now}$ raises a question of the dependence of a(t) and the other quantities on the current time. To remove that impression we will express the previous results in terms of a different base time T' instead of t_{now}.

$$a(t, T) = a'(t', T') = (t/T')^{g' + h't'} \tag{3.13}$$

where

$$t' = tT/T' \tag{3.14}$$
$$g' = g$$
$$h' = hT/T'$$

Although the parameterization appears different the physical results are the same.[15]

3.9 Why is the Fit So Successful?

The phenomenological fit in eq. 3.1 appears to be a successful fit to the growth of the universe. It raises the question: Why is it so successful? It matches the Big Bang model at the beginning. It accounts for the observed rise in H(t). It shows an initial

---

[15] In that regard it is analogous to the renormalization group analysis of coupling constants with "coupling constants" $h = h(t_{now})$, $h' = h(t_m)$, and the times t_{now} and t_m being analogous to energy scales. See chapter 5.

decline in H(t) after the end of the Big Bang epoch. It matches our expectation that the rise in H(t) is the result of an energy inflow from the Megaverse.

In this author's view its success is probably attributable to renormalization group self-similarity.[16] The evolution of universes may have laws of growth from a Big Bang metastate that are based on self-similar growth along the lines of the Renormalization group. Chapter 5 discusses a close analogy with quantitative similarities between electron vacuum polarization and the form of the growth of the universe.

Note: all logarithms in the following plots are to base 10.

---

[16] The self-similarity is based on the principle that the physics of the evolution of universes should be independent of the choice of the present time t_{now}. A change in t_{now} should result in a corresponding change in parameters—the essence of the Renormalization group.

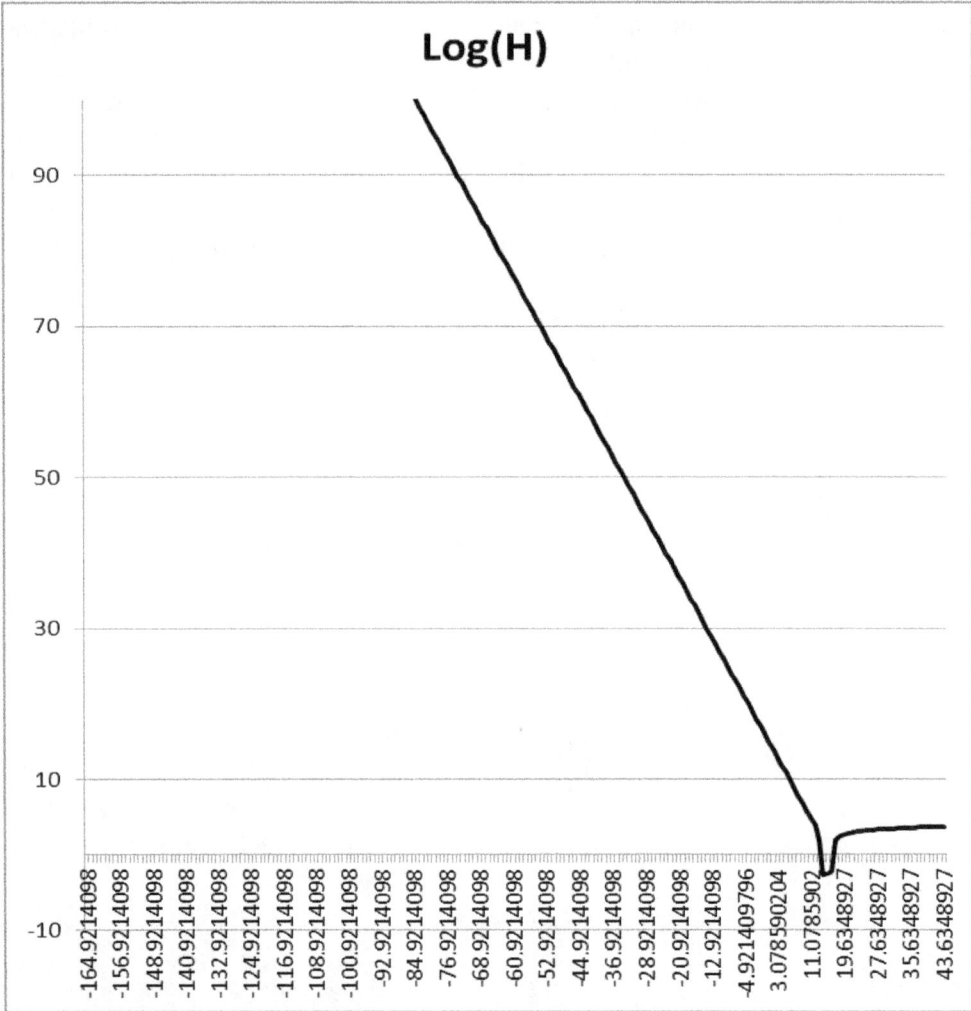

Figure 3.1. Log H(t) plotted vs. log t from the Big Bang metastate to the present. Note the dip below zero of H(t). Also note H(t) rises after the dip as measured experimentally.

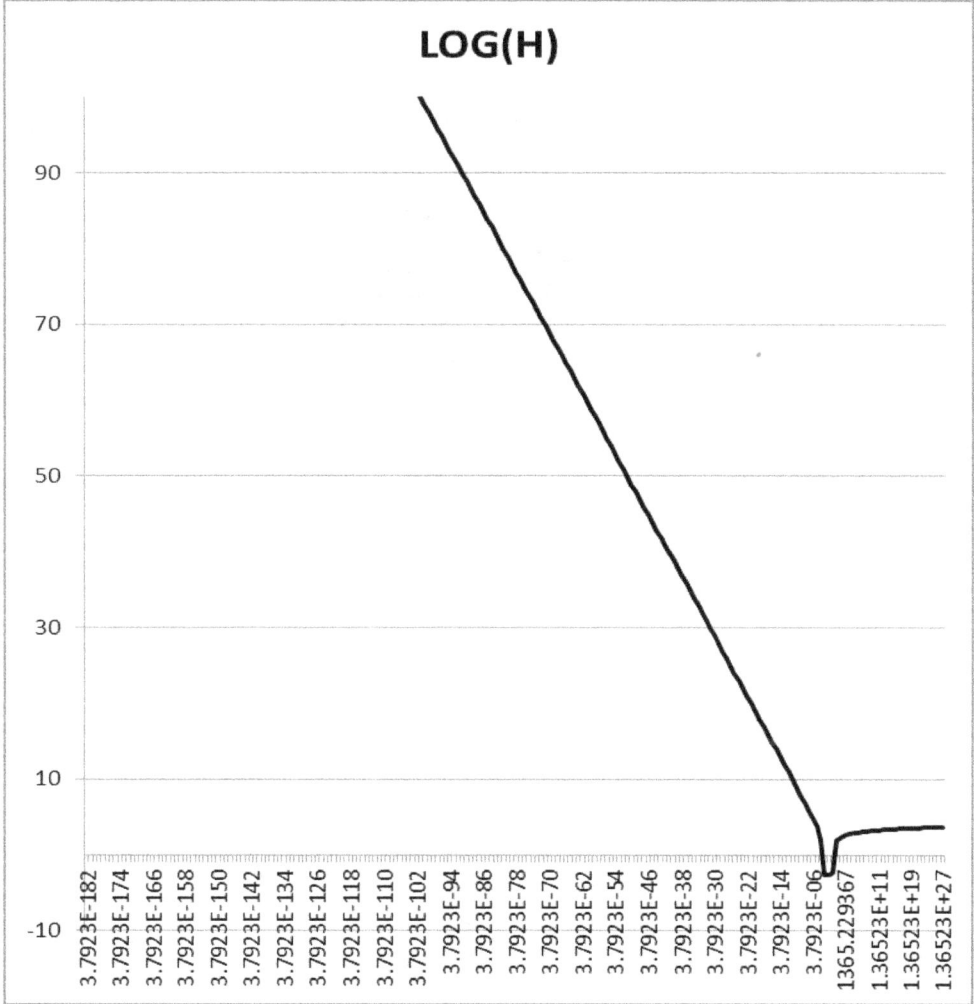

Figure 3.2. Log H(t) plotted vs. time in Gyr' from the Big Bang metastate to the present. Note the dip below zero of H(t). Also note H(t) rises after the dip as measured experimentally. The number following "E" is a power of ten.

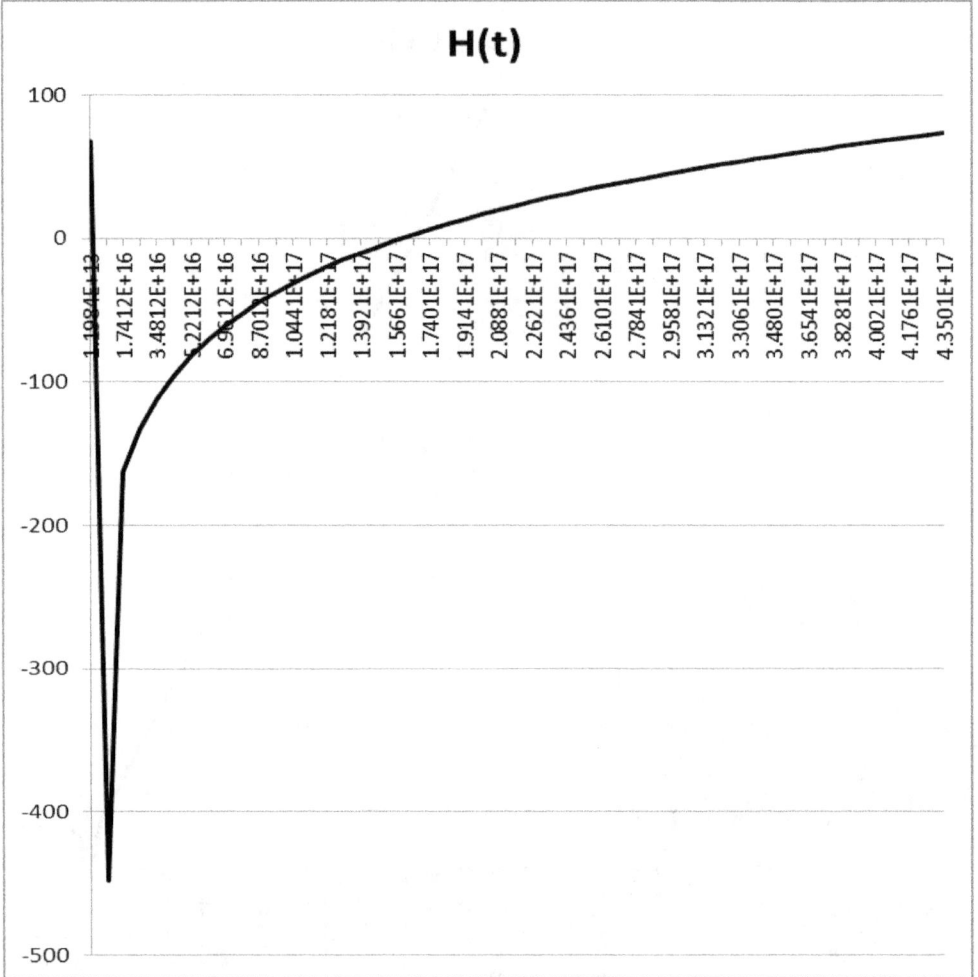

Figure 3.3. H(t) plotted vs. time in Gyr's from the year 380,000 to the present. Note the pronounced dip below zero of H(t). Also note H(t) rises after the dip as measured experimentally. The low point of the dip occurs at t = 8.71 × 10^{13} s.

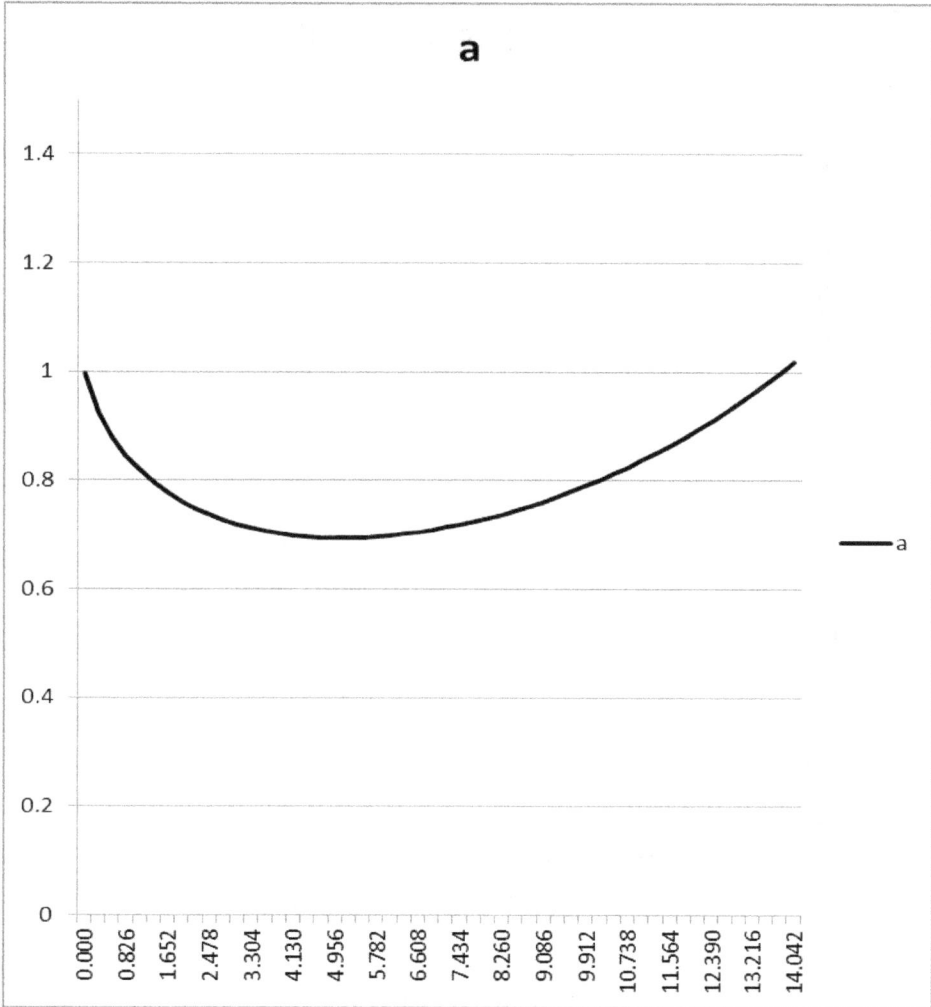

Figure 3.4. a(t) plotted vs. time in Gyr's from the year 380,000 to the present. Note the dip in a(t) (a smaller universe?) and then rises again to its current size.

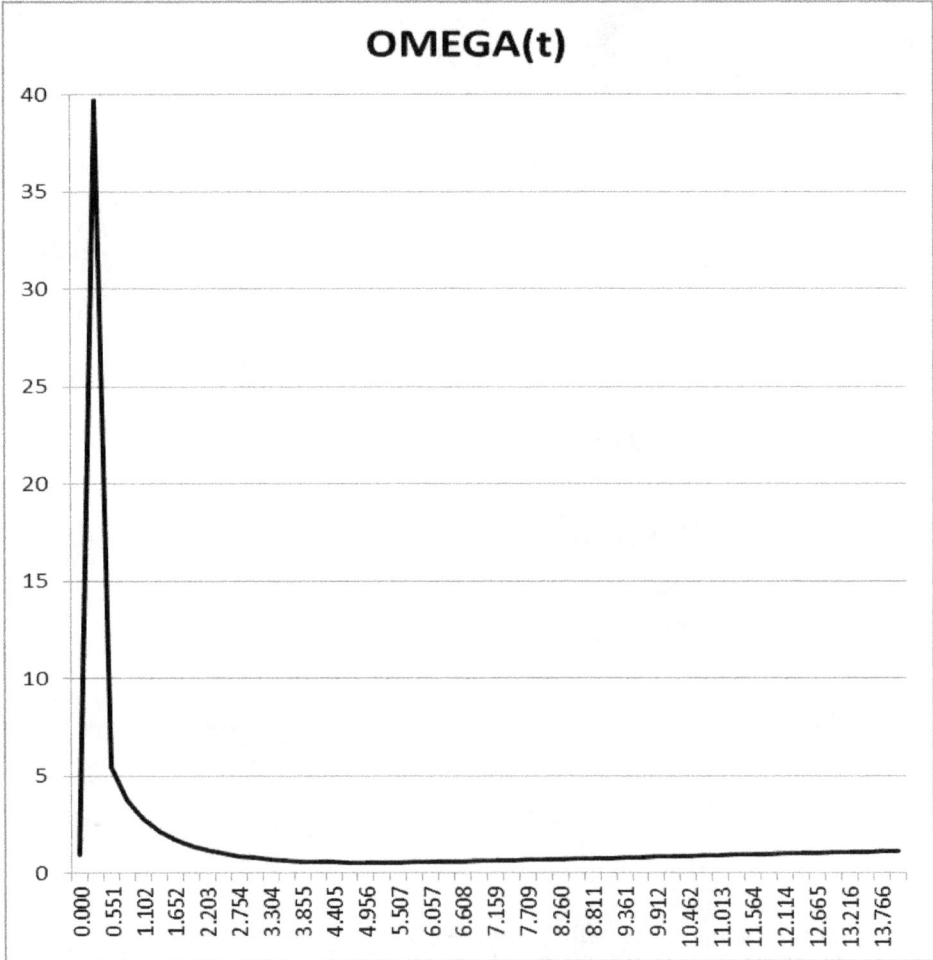

Figure 3.5. $\Omega_T(t)$ plotted vs. time in Gyr's from the year 380,000 to the present. Note the peak suggesting an influx into the universe at the transition to the matter-dominated phase, and then a decline followed by a raise.

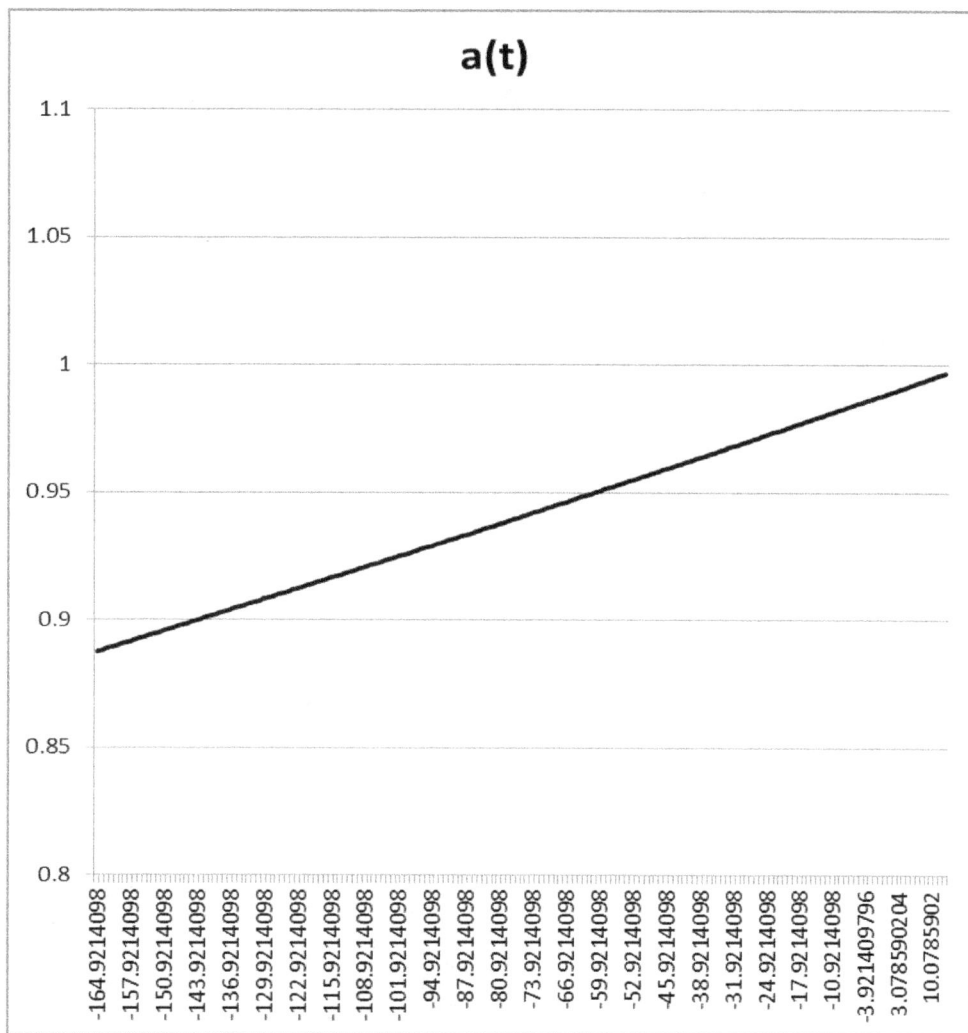

Figure 3.6. a(t) plotted vs. log time in seconds from the Big Bang metastate to to t = 13.65 Gyr. Note the dip in a(t) (a smaller universe?) after 380,000 yrs followed by a rise.

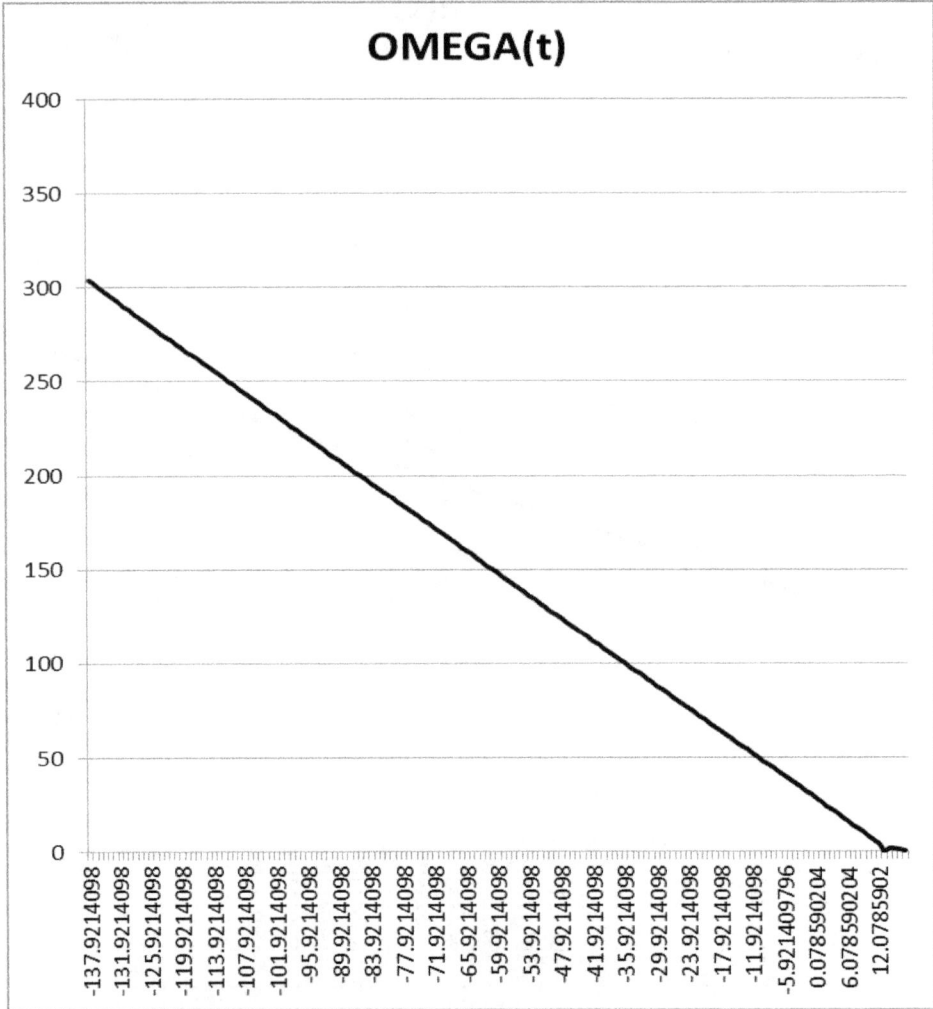

Figure 3.7. Log $\Omega_T(t)$ plotted vs. log time in seconds from 3.8×10^{-155} Gyr to the 13.65 Gyr. Note the scarcely visible fluctuation in dip in $\Omega_T(t)$ (a smaller universe?) after 380,000 yrs. See Fig. 3.5 for a detailed view.

4. Further Plots of Universal Scale Factor Quantities

The study of the implications of our universal scale factor formulation for the evolution of the universe, and particularly for the unforeseen Big Dip that we found, requires a detailed look at the various related quantities. In this chapter we plot quantities in eqs. 3,7 through 3.11. The plots tend to support the validity of the formulation. Some display further indications of the Big Dip that occurred after the radiation-matter transition.

The complete universe "life" picture presented in the following plots is consistent with a declining pressure from the Big Bang phase, a declining density, and a declining pressure consistent with our physical expectations. The Big Dip and subsequent rises would appear to be due to a sharp influx of energy (Fig. 3.5) from the Megaverse..

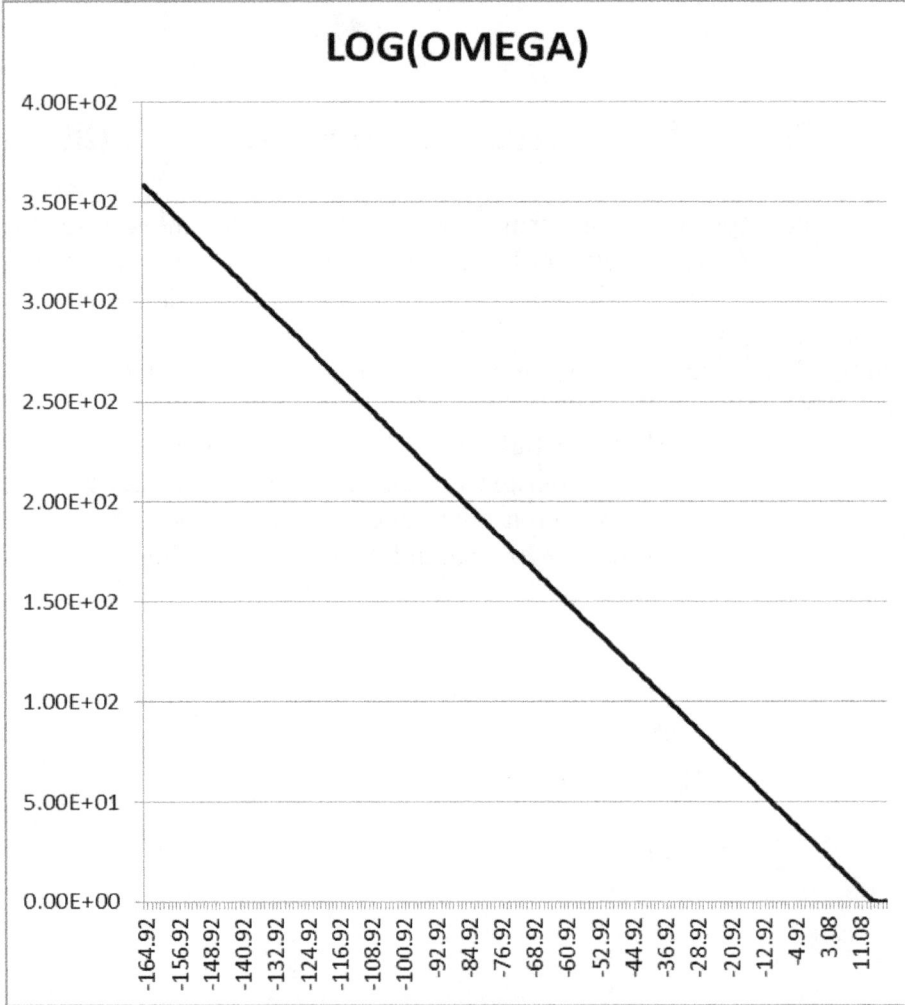

Figure 4.1. Log_{10} $\Omega_T(t)$ plotted vs. $\log_{10}$ t from the Big Bang metastate to the present. Note the "dip" below zero of $\Omega_T(t)$ at $\log_{10}$ t = 11.

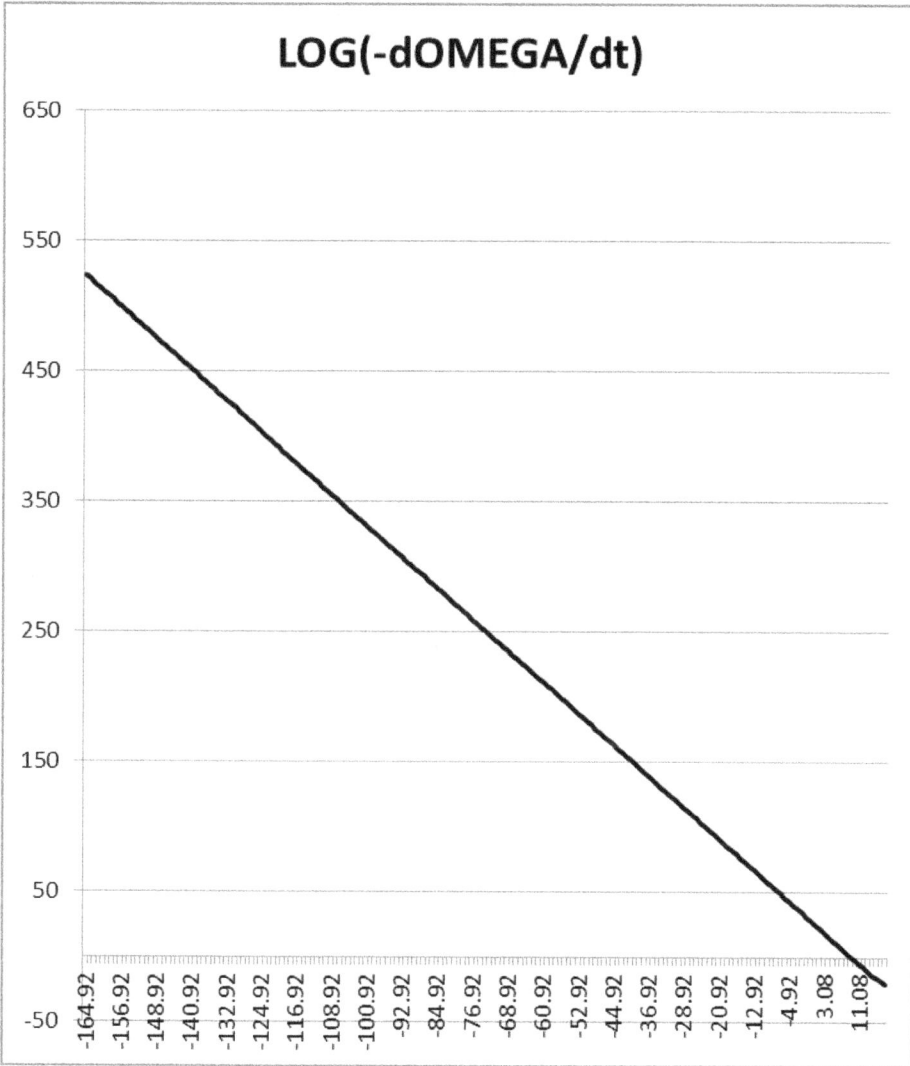

Figure 4.2. Log_{10} ($-d\Omega_T(t)/dt$) plotted vs. log_{10} t from the Big Bang metastate to the present. Note the dip below zero at log_{10} t = 11.

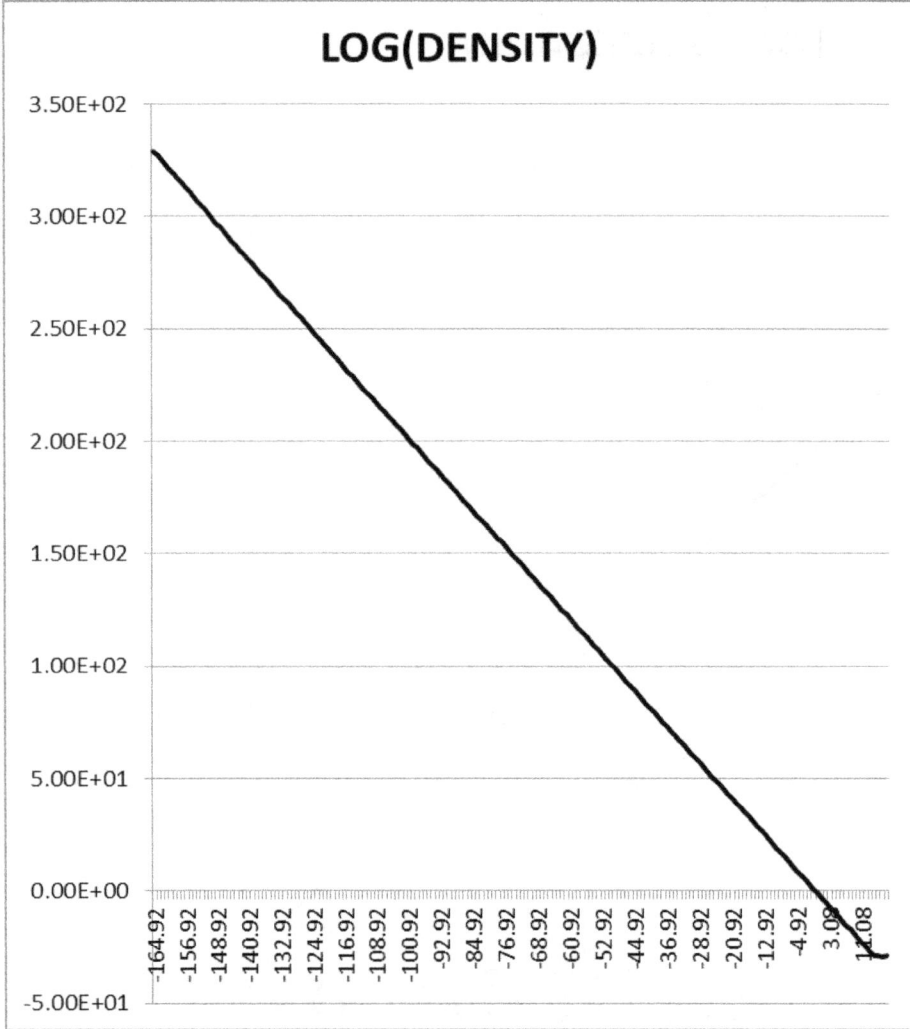

Figure 4.3. Density. $\text{Log}_{10}\,\rho(t)$ plotted vs. $\log_{10} t$ from the Big Bang metastate to the present. Note the dip below zero at $\log_{10} t = 3$.

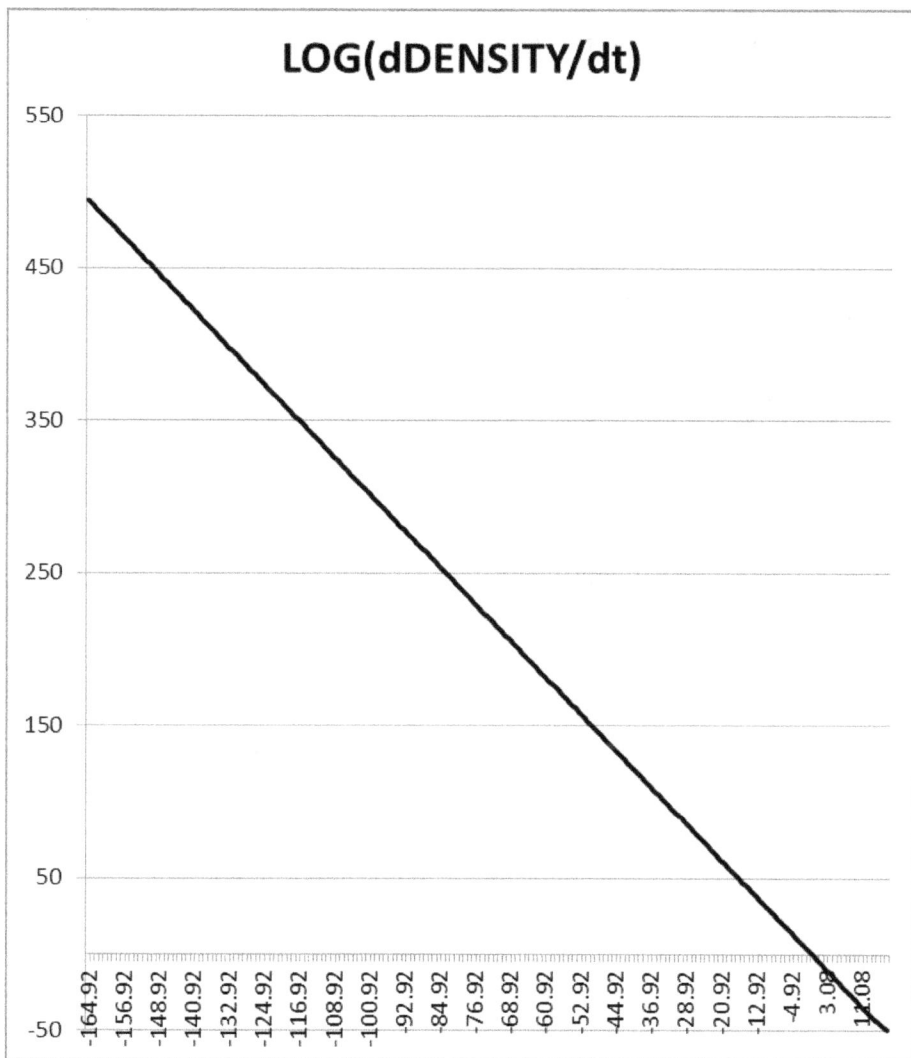

Figure 4.4. Log_{10} $d\rho(t)/dt$ plotted vs. log_{10} t from the Big Bang metastate to the present. Note the dip below zero at log_{10} t = 3.

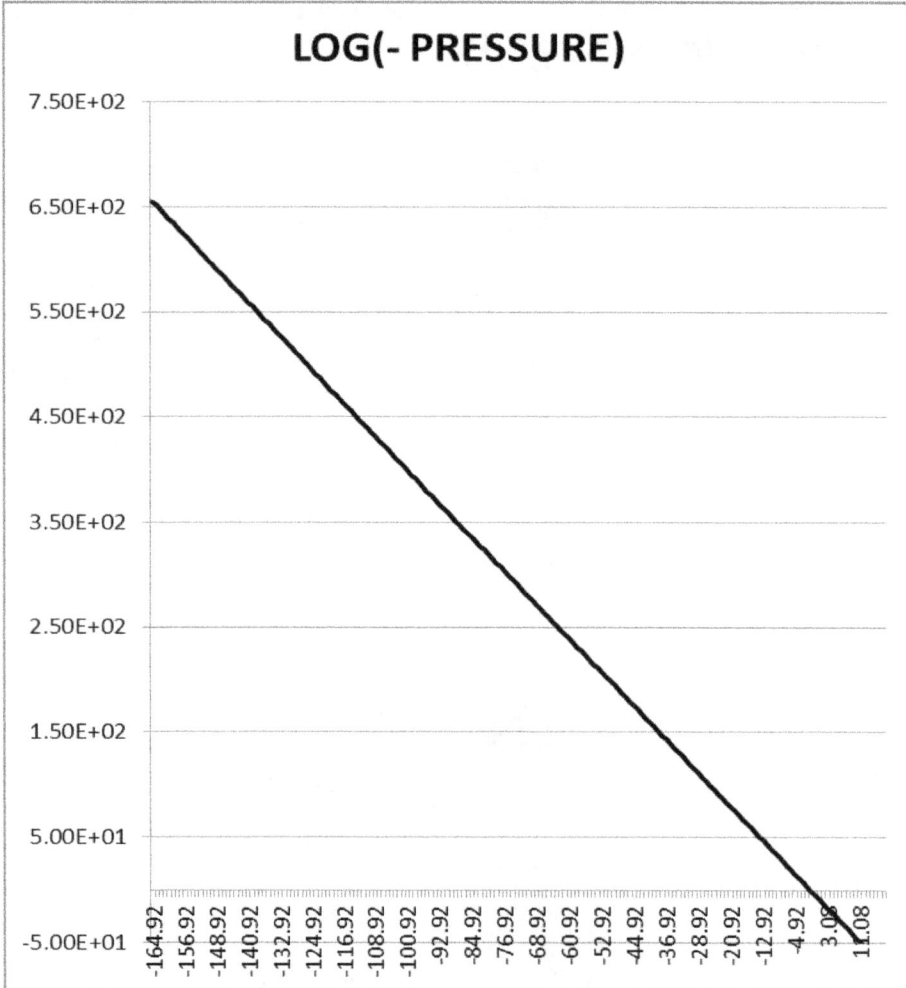

Figure 4.5. Pressure.. $\log_{10}$ (-p(t)) plotted vs. $\log_{10}$ t from the Big Bang metastate to the present. Note the dip below zero at $\log_{10}$ t = 3. The pressure is negative indicating a pressure for expansion. The dip below zero indicates relatively very low pressure.

5. Fundamental Justification of the Universal Scale Factor

In the preceding chapters it has become evident that the universal scale factor specified by eq. 3.1 leads to a physically reasonable structure for the life history of the universe. The question that immediately arises is the cause of this form for a(t). How does it work so well for both extremely early times near the Big Bang and for recent times as well?

Its form

$$a(t) = (t/t_{now})^{g + ht} \tag{3.1}$$

can be expressed as

$$a(t) = (t/T)^{g + t/T} \tag{3.1}$$

where T is the "latest" time using eqs. 3.1a and 3.13. The value of the exponent g is

$$g = 0.000283993 \tag{5.1}$$

For early times near the Big Bang time of the order of 10^{-165} s we can approximate

$$a(t) \cong (t/T)^g \tag{5.2}$$

5.1 Significance of the Value of g

At first glance the value of g is an arbitrary constant set through the values of the Hubble Constant at t = 380,000 years and t_{now}. However the value of g is remarkably similar to the value of a renormalization exponent in the author's calculation of the Fine Structure Constant in 1973 and in section 5.5.3 of Blaha (2019b). The value of the Fine Structure Constant was based entirely on vacuum polarization in massless Quantum

Electrodynamics (QED) – a vacuum effect of electromagnetism. Although astrophysicists do not think of the Big Bang and the expansion of the universe as a vacuum effect, it is clear from the plots shown earlier that universe growth is dependent on its energy density and an influx (appearance) of energy from somewhere (the Megaverse in our view). Thus the growth of the universe is directly dependent on the vacuum (of the Megaverse).

In massless QED we found that the vacuum polarization had the form:[17]

$$f(\alpha)(p^2/\Lambda^2)^{g_{QED}} \tag{5.3}$$

where $f(\alpha)$ is the "eigenvalue function" for the Fine Structure Constant[18] of the Johnson-Baker-Willey model of massless QED, p is the momentum, and Λ is the ultraviolet cutoff. The value of g_{QED} that corresponded to the Fine Structure Constant was[19]

$$g_{QED} = -0.000580537 \tag{5.4}$$

and the Fine Structure Constant was correctly found (well within experimental limits) to be

$$\alpha_{calculated}(g_{QED}) = 0.007297354 \tag{5.5}$$

To compare our g with g_{QED} we could simply reexpress eq. 5.2 as

$$a(p) \cong (p^2/\Lambda^2)^{-g} \tag{5.2a}$$

setting $t = p^{-2}$ and $T = \Lambda^{-2}$. As a result since

$$g = 0.000283993 \tag{5.6}$$

we find

---

[17] Eq. 12 in S. Blaha, Phys Rev **D9**, 2246 (1973).
[18] The author calculated $\alpha = 1/137\ldots$ exactly (within experimental limits) in Blaha (2019a) and (2019b).
[19] Section 5.5.3 of Blaha (2019b).

$$-g \cong \frac{1}{2} \, g_{QED} \qquad (5.7)$$

Except for the factor of two we find a remarkable numerical coincidence comparing electron vacuum polarization with the universe expanding in the vacuum (which we take to be the vacuum of the Megaverse).

5.1.1 A New Unknown Interaction determining Universe Evolution

A deeper way of relating eqs. 5.2 and 5.2a is to fourier transform the ultra-small scale factor approximation eq. 5.2 to "momentum" space:

$$D(p) = (1/2\pi) \int_0^\infty dt \, \exp(-ip^4 t) \, (t/T)^g \qquad (5.8)$$

$$= \frac{\text{const.}}{P^4} \, (p^2/\Lambda^2)^{-2g} \qquad (5.8a)$$

where[20]

$$1/T = \Lambda^4$$

with Λ being a "momentum space" cutoff mass. The exponent agrees well with QED:

$$-2g = -0.000567986 \cong g_{QED} \qquad (5.8b)$$

5.1.2 A New Interaction?

Examining eq. 5.8a in more detail we find it can be viewed as a renormalized propagator for a vector (or spin 2) gauge theory where the bare propagator is

$$1/p^4$$

---

[20] The p^4 factor in the exponent was chosen to compare the QED exponent with the $a(t)$ exponent for ultra-short times.

This type of propagator is used in Blaha (2018e) as the strong interaction gluon propagator[21] that gives a confining r potential for quarks, and in linearized gravitation theory[22] to generate MOND-like potentials.

Thus one can view the fourier transform of a(t) at ultra-short times as a renormalized propagator D(p) (modulo a $g_{\mu\nu}$ or a tensor polynomial) for an interaction that fixes the time evolution of the universe's density, and thus a(t) and the Hubble Constant. This interaction is at least part of the large Ω_T seen in Fig. 3.5 and other figures.

The possible new interaction has a coupling constant α_U (assuming our general formulation of vacuum polarization in Blaha (2019b) is applicable) that is given by eq. 5.13 below. Its value is about half of the QED Fine Structure Constant α. This value suggests that the new interaction, if it exists, is QED-like and not like that of Gravity.

By manipulating the energy density of the universe the interaction vacuum polarizes the Megaverse? vacuum guiding the growth of the universe. Chapter 6 considers the creation of the universe and an anti-universe pair as a vacuum event. It suggests that we can view a universe as a particle.

Considering the entire life of the universe as a whole, we can view the growth the universe as a Megaverse vacuum polarization event just as an electron can be view as surrounded by a polarized vacuum.[23]

5.2 Explaining the Factor of Two in the Exponents

The factor of two (eq. 5.7) requires explanation if there is more than a simplistic coincidence in the numeric values of the g's. We will use the expressions for the coupling constants of non-abelian gauge theories in Blaha (2019b) as the starting point. The QED Fine Structure Constant satisfied

$$(\alpha/2\pi) = [g_{QED}A_4 - (4 + 2g_{QED})A_2]/(A_4A_1 - A_2A_3)$$

---

[21] See eq. 24.18 in chapter 24 of Blaha (2018e).
[22] See eq. 23.19 in chapter 23 of Blaha (2018e).
[23] We thus view the entire time evolution of the universe as a whole. Normally one views time as increasing. Feynman suggested we could also view time as flowing backward. In this section we freeze the life history of the universe as a static event rather like a time lapse picture of a flower's growth that we sometimes see.

where A_1, A_2, A_3, and A_4 are functions of g_{QED}. We found a similar equation yielded the coupling constants of SU(2), SU(3), and SU(4):

$$c_G^{-1} = [(11/3)C_{ad} - 2C_f/3]/(16\pi)^3 \qquad (6.1)$$
$$(\alpha_G/2\pi) = c_G^{-1}[gA_4 - (4 + 2g)A_2]/(A_4A_1 - A_2A_3) \qquad (7.2)$$

where C_{ad} is the dimension of the fundamental representation of the non-abelian group and C_f is the number of fermions (fermion flavor) of the interaction.with equation numbers from Blaha (2019b).

Since the Lorentz group of QED has a 4-dimensional fundamental representation, we can set

$$c_{GQED}^{-1} = C_{ad}/4 \qquad (5.9)$$

with $C_{ad} = 4$ thus obtaining eq. 5.8 above.

Now a(t) is complex-valued as shown in Blaha (2019c) although the imaginary part of a(t) $\approx 10^{-93}$ is small. Complex-valued rotations of a(t) rotate the real and imaginary parts into each other. Thus $C_{ad} = 2$ and

$$c_{GU}^{-1} = \frac{1}{2} \qquad (5.10)$$

for the universe U. Therefore eq. 7.2 above becomes

$$(\alpha_{GU}/2\pi) = c_{GU}^{-1}[g_UA_4 - (4 + 2g_U)A_2]/(A_4A_1 - A_2A_3) \qquad (5.11)$$
$$= \frac{1}{2}[g_UA_4 - (4 + 2g_U)A_2]/(A_4A_1 - A_2A_3)$$

where

$$g_U = 0.000283993 \cong -\frac{1}{2} g_{QED} \qquad (5.12)$$

The corresponding coupling constant α_U has the value

$$\alpha_U = 0.003569265 \qquad (5.13)$$

which is about half of the QED Fine Structure Coupling Constant. It may indicate the existence of an interaction that may have a role in the evolution of the universe that may account for eq. 5.1.

5.3 Using g_{QED} in the Universal Scale Factor

Suppose we use

$$a_U(t) = (t/t_{now})^{g_U + t/t_{now}} \tag{5.14}$$

where we set

$$g_U = -\tfrac{1}{2}\, g_{QED} = 0.000290269 \tag{5.15}$$

Then if we calculate H(380,000 yrs) and H(t_{now}) we find

$$H(380,000 \text{ yrs}) = 77.45$$

$$H(t_{now}) = 74.47$$

in serious disagreement with the currently accepted values.

5.4 Fundamental Basis for the Universal Scale Factor

This chapter strongly suggests that the universal scale factor that we have found may have a deep underlying basis.

6. Universe Particle Dynamics

The origin of our universe and other universes is a weighty question that has been considered in a number of theories. Our view is that the existence of the Megaverse, and the locality of universes (in the one case of which we are certain), suggest the Megaverse is the "platform" of a type of particle physics in which universes play the role of particles. There are clear points of difference: particles have a fixed mass, particles have well-defined quantum numbers, and so on. Nevertheless we can consider universes to be variable mass particles, *universe particles*, with a dynamics that is primarily based on gravitation.

This chapter describes aspects of the interactions of universe particles due to gravitation, baryon number forces, and collisions between universes. It also the describes the genesis of universes due to vacuum fluctuations, the fission of universes, and the internal distortion of universes due to acceleration and the presence of 'nearby' universes..

6.1 The Internal Distortion of Universes

In the absence of external forces universes are considered to be uniform in the large. However the acceleration of a universe in the Megaverse can distort the universe. Also the existence of a nearby universe(s) could cause the uniformity of a universe to be lost due to gravitation and baryonic forces.[24]

6.1.1 Impact of Universe Particle Acceleration – Lopsided Internal Structure of Universe

Universes can accelerate within the Megaverse due to external Megaverse forces. Universe acceleration should be detectable within a universe as "lopsidedness."

---

[24] The baryonic forces, the Baryonic force and the Dark Baryonic force, on a universe are large due to their additivity in our universe and nearby universes.

There would be a shift of parts of the universe opposite to the direction of acceleration resulting in a difference in the features of the universe "in front" compared to those "in back" – an acceleration effect just as one sees when a jet accelerates.

Interestingly new data from the Planck observatory of the European Space Agency confirms and extends earlier data from NASA's WMAP observatory that one side of the universe appears different from the other side. There are temperature differences and mass distribution differences – just as one might expect if the universe were accelerating as a unit.

Thus we see the beginning of data suggesting our universe may be accelerating through the Megaverse. Some Planck observatory scientists have suggested their data is a preliminary indication of the Megaverse.

6.1.2 Impact of External Forces on Universe Structure

The presence of a nearby universe could cause a universe to lose its large scale uniformity and the mass-energy of the universe to drift over time to the 'nearby' side of the universe due to gravitation and baryonic forces.

6.2 Universes in Collision

We can assume that the dynamics of universes in collision will be analogous to that of galaxies in collision since gravity is a dominant force in both cases. Colliding galaxies have often been observed. Their dynamics should provide guidance for the case of universes in collision.[25]

It is clear in the case of colliding galaxies, and of colliding large nuclei (gold and lead typically) that there are several types of collisions with differing results. Similarly, the types of universe collisions can be qualitatively classified as:

1. Clean collisions in which universes nudge each other but retain their identity. These are extreme peripheral collisions. If the universes overlap slightly then the

---

[25] The high energy collision of atomic nuclei at Brookhaven, CERN and other laboratories also is analogous in overall detail with universes in collision.

typically spherical symmetry of the universes may become distorted and they may become lopsided.[26]

2. Peripheral collisions in which the universes retain their identity but are connected by a trailing string of mass-energy. Eventually the string breaks and the universes separate. Subsequently the pieces of trailing string in each universe contract due to their universe's gravitational effects and perhaps form new "bubble" universes.

3. Two universes can collide and produce multiple universes.

4. Two universes can collide in a "central" collision and amalgamate into one universe. They can intermix with both the baryonic, gauge, and gravitational forces causing a redistribution of their masses. They may separate afterwards or may coalesce into a single universe. One result of this may be lopsided universes. Our universe appears to be lopsided. Some cosmologists believe this is due to a near collision of our universe with another shortly after the Big Bang.

6.3 Creation of Universes through Gauge Field Fluctuations

One of the most exciting questions in Cosmology is the origin of our universe. The conventional view is that it originated in a Big Bang from an infinitesimal point in space. The source of the Big Bang and the prior state of the Megaverse, if there was one, is the subject of much speculation. Based on the particle interpretation of the Wheeler-DeWitt equation, the possibility of a baryonic force strongly supported by conservation of baryon number, and the Megaverse concept, it is reasonable to consider the possibility that the universe originated in a vacuum fluctuation.

In this case there would be two Big Bangs one for our universe and one for an anti-universe. One would expect that they would have opposite corresponding features:

---

[26] The Wilkinson Microwave Anisotropy Probe (WMAP) and the Planck European Space Agency satellite have been accumulating data since 2001 that suggests the universe may be lopsided with hot and cold spots on opposite sides of the universe differing from those on the other side being hotter and colder respectively—*perhaps the result of a collision when the universe was young.*

one with baryon dominance – one with anti-baryon dominance, and one left-handed – one right-handed.

Our formulation of universe particle theory allows for the generation of a universe particle and anti-particle as a vacuum fluctuation. We view a universe particle as having a substantial excess of baryons, N, as we see in our universe. Its anti-universe at the time of creation (the Big Bang point) is its "mirror image" having the "same" number of anti-baryons (baryon number –N) so that baryon number is conserved by the fluctuation event. Thus the excesses of one universe are compensated by the excesses of the other.

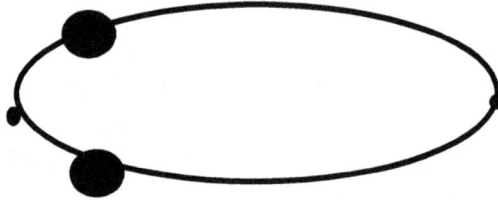

Figure 6.1. Generation of a universe – anti-universe pair as a vacuum fluctuation.

The small value of the coupling constant should lead to an extremely long lifetime for the universes generated by the fluctuation. Thus the 45 billion year life of our universe is not unreasonable. The probability of the creation of universes by vacuum fluctuations should be correspondingly small.

The sizes of the created universe and antiuniverse should be very small just as Big Bang theories of our universe suppose.

6.4 Fission of Universes

Under certain circumstances the distribution of matter in the universe may lead to the fission of the universe into two separate universes. Our theory supports this possibility for universe particles. The detailed mechanism of the fission process is not specified by the model.

6.4.1 Fission of Normal Universes

The fission of universe particles in our universe particle model is depicted in the Feynman diagram in Fig. 6.2.

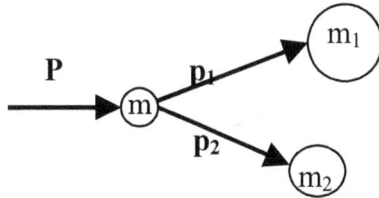

Figure 6.2. Fission of a universe particle into two universe particles.

The sum of the masses of the output universe particles is usually less than the original universe particle mass. However if the fission takes a long time and the masses are time dependent then the produced universe particles combined masses may exceed the original universe's mass.

6.4.2 Tachyon Universe Particle Fission to More Massive Universe Particles

In Blaha (2007a) we showed that a tachyonic (faster than light) particle could fission into particles of larger mass through the conversion of momentum into mass. In this section we show that a tachyonic universe particle may fission into two more massive universe particles.[27] This phenomenon is of particular interest because it enables tachyonic universes to spawn in a new novel way not previously considered in discussions of the origin of universes.

A simple model lagrangian[28] for a tachyonic universe particle is

$$\mathcal{L}_{\parallel} = \psi_T^S(Y(y))[\gamma^{\mu}\partial/\partial y^{\mu} - e_B\gamma^{\mu}B_{u\mu}(Y(y)) - m(t)]\psi_T(Y(y)) - \tfrac{1}{4}F_{Bu}^{\mu\nu}(Y(y))F_{Bu\mu\nu}(Y(y))$$
$$- \tfrac{1}{4}F_u^{\mu\nu}(y)F_{u\mu\nu}(y)$$

---

[27] We will use the term mass here to denote mass-energy. Since we identified mass as a multiple of area earlier, the comments here would appear to apply to universe area as well.
[28] See Blaha (2018e) and earlier books by the author for a detailed discussion.

We assume m(t) is constant.

When a particle or a universe particle fissions (decays) one normally expects that the masses of the particles or universe particles produced by the decay to be smaller than the mass of the original particle or nucleus. In the case of tachyonic (faster-than-light) elementary particles, or universe particles, a much different possibility is present: a tachyon universe can decay into heavier tachyons (perhaps through a distortion of the universe internally into two 'lumps'.) We consider the specific case of a tachyon universe particle decaying into two universe particles whose total mass is greater than the original. (See Fig. 6.3.)

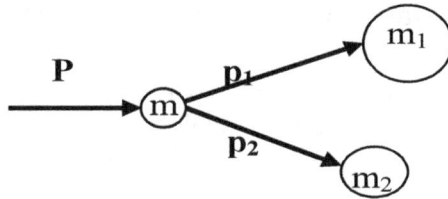

Figure 6.3. Two universe particle decay of a tachyon universe particle.

We will assume the initial tachyon universe particle has zero energy ($p^D = 0$) and thus the tachyons universe particles emerging from the decay also have total universe particle energy zero. The analysis is based on conservation of total universe energy and momentum in Megaverse space outside of universes. The below discussion applies to D-dimensional space with (D – 1)-dimensional spatial coordinates.

Momentum conservation implies

$$\mathbf{P} = \mathbf{p}_1 + \mathbf{p}_2$$

Since all energies are zero

$$(c P)^2 = (c \mathbf{P})^2 = m^2$$
$$(c p_1)^2 = (c \mathbf{p}_1)^2 = m_1{}^2$$
$$(c p_2)^2 = (c \mathbf{p}_2)^2 = m_2{}^2$$

where $P = |\mathbf{P}|$, $p_1 = |\mathbf{p_1}|$, and $p_2 = |\mathbf{p_2}|$. If we now square the above equation for $\mathbf{P}$ and then use the above three equations we obtain

$$m^2 = m_1^2 + m_2^2 + 2m_1m_2 \cos \theta$$

where θ is the opening angle between the emerging universe particles momenta of $\mathbf{p_1}$ and $\mathbf{p_2}$. There are a number of interesting cases:

Case $\theta = 0$:

$$m = m_1 + m_2$$

The masses of the outgoing universe particles sum to the mass of the original tachyon universe particle.

Case $\theta = \pi/2$:

$$m^2 = m_1^2 + m_2^2$$

The masses of each outgoing universe particle tachyon are less than the mass of the original tachyon universe particle.

Case $\theta = \pi$:

$$m^2 = (m_1 - m_2)^2$$

In this case either $m_1 > m$ or $m_2 > m$. Thus one of the outgoing tachyon universe particles has a greater mass than the original tachyon universe particle. Mass is effectively created from the spatial momentum of the initial universe particle. This process is the inverse of normal particle and universe particle fission where the sum of the outgoing masses is always less than the original particle's mass and the difference is mass converted into energy in the form of additional photons.

This last case, where one of the outgoing universe particles is more massive than the original universe particle, is not just for $\theta = \pi$. Since

$$\cos \theta = (m^2 - m_1{}^2 - m_2{}^2)/(2m_1m_2)$$

we see that the sum of the outgoing universe particle masses is always greater than the original tachyon universe particle *mass (except when $\theta = 0$)* since

$$\cos \theta = 1 + [m^2 - (m_1 + m_2)^2]/(2m_1m_2) \leq 1$$

and thus

$$[m^2 - (m_1 + m_2)^2]/(2m_1m_2) \leq 0$$

Note $m = m_1 + m_2$ only if $\theta = 0$.

Since we can transform the above discussion to the case of universe particle tachyons having non-zero Megaverse energy using an ordinary D-dimensional Lorentz transformation, the discussion in this subsection is general.

We therefore conclude that when a tachyon universe particle decays into two tachyon universe particles the sum of the masses of the produced tachyon universe particles is greater than the mass of the original tachyon universe particle except if the angle between the momenta of the produced tachyon universe particles is zero. In that case the sum of the masses of the produced tachyon equals the mass of the original tachyon universe particle and the produced universe particles overlap.

Appendix A. Some Physical Constants Used in Calculating Numerical Expressions

Some physical constants that we found to be of use in the evaluation of expressions are (assuming units with $c = \hbar = 1$):

$$G \equiv 7.39 \times 10^{-29} \text{ gm}^{-1} \text{ cm} \qquad \text{(A.1a)}$$
$$G \equiv 2.6 \times 10^{-66} \text{ cm}^2 \qquad \text{(A.1b)}$$
$$G \equiv 2.91 \times 10^{-87} \text{ s}^2 \qquad \text{(A.1c)}$$

$$H_0 \equiv 2.133 \times 10^{-33} \text{h ev} \qquad \text{(A.2a)}$$
$$H_0 \equiv 1.08 \times 10^{-28} \text{h cm}^{-1} \qquad \text{(A.2b)}$$
$$H_0 \equiv 3.24 \times 10^{-18} \text{h s}^{-1} \qquad \text{(A.2c)}$$
$$h = .689 \qquad \text{(A.2d)}$$
$$H_0 = 100 \text{ h km s}^{-1} \text{ Mpc}^{-1} \qquad \text{(A.2e)}$$
$$1 \text{ s}^{-1} = 3.0864 \times 10^{-20} \text{ km s}^{-1} \text{ Mpc}^{-1} \qquad \text{(A.2f)}$$

$$GH_0 \equiv 7.98 \times 10^{-57} \text{h gm}^{-1} \qquad \text{(A.3)}$$

$$\rho_{\text{crit}} \equiv 1.88 \times 10^{-29} \text{h}^2 \text{ gm cm}^{-3} \qquad \text{(A.4)}$$

$$M_{\text{Planck}} \equiv 1.22 \times 10^{28} \text{ ev} \qquad \text{(A.5a)}$$
$$M_{\text{Planck}} \equiv 2.18 \times 10^{-5} \text{ g} \qquad \text{(A.5b)}$$

$$M_{\text{Planck}} \equiv 6.20 \times 10^{32} \text{ cm}^{-1} \qquad \text{(A.6a)}$$
$$\text{Planck Length} = M_{\text{Planck}}^{-1} \equiv 1.61 \times 10^{-33} \text{ cm} \qquad \text{(A.6b)}$$

$$M_{Planck} \equiv 1.85 \times 10^{43} \text{ s}^{-1} \tag{A.7a}$$

$$\text{Planck time} = M_{Planck}^{-1} \equiv 5.41 \times 10^{-44} \text{ s} \tag{A.7b}$$

$$1 \text{ eV} \equiv 5.08 \times 10^{4} \text{ cm}^{-1} \tag{A.8a}$$

$$1 \text{ eV} \equiv 1.52 \times 10^{15} \text{ s}^{-1} \tag{A.8b}$$

$$1 \text{ eV}/c^{2} \equiv 1.783 \times 10^{-33} \text{ g} \tag{A.8c}$$

$$1 \text{ g} \equiv 2.85 \times 10^{37} \text{ cm}^{-1} \tag{A.8d}$$

$$\kappa \equiv 4.38 \text{ °K}^{-1} \text{ cm}^{-1} \tag{A.9}$$

$$\kappa \equiv 1.31 \times 10^{11} \text{ °K}^{-1} \text{ s}^{-1} \tag{A.10}$$

$$\kappa \equiv 8.62 \times 10^{-5} \text{ ev °K}^{-1} \tag{A.11}$$

$$1 \text{ Gyr} = 3.16 \times 10^{16} \text{ s} \tag{A.12}$$

where κ is Boltzmann's constant.

Appendix B. Defined Quantities in the Text

This appendix lists constants defined by the Particle Data Group and in the text.

Constants based on M. Tanabashi *et al op. cit.*

h = Hubble Constant = 0.678(9)

ρ_{crit} = Critical density = $1.87840(9) \, h^2 \times 10^{-29}$ g/cm^{-3}

Ω_Λ = Dark Energy density/ρ_{cr} = ρ_{de}/ρ_{cr} = 0.692 ± 0.012

$\Omega_d = \Omega_c$ = Cold Dark matter density/ρ_{cr} = ρ_c/ρ_{cr} = $0.1186(20) \, h^{-2}$

Ω_b = Baryon density/ρ_{cr} = ρ_b/ρ_{cr} = $0.02226 \, h^{-2}$

$\Omega_m = \Omega_b + \Omega_c$

   = pressureless Matter density/ρ_{cr} = ρ_m/ρ_{cr} = 0.308 ± 0.012

$\Omega_{mtot} = \Omega_m + \Omega_\Lambda = 1.000 \pm 0.024$

$t_0 = t_{now}$ = Age of universe = 13.80 ± 0.04 Gyr

$r_{universe}(t_{now})$ = visible radius of the universe = 4.314×10^{28} cm

(13.1.3)

Ω_γ = radiation density/ρ_{cr} = ρ_γ/ρ_{cr} = $2.473 h^{-2} \times 10^{-5} \, (T/2.7255)^4 h^{-2}$

   = 5.38×10^{-5}

(13.1.4)

$k = 5.56 \times 10^{-57}$ cm^{-2}

Constants defined in the text:

$$C_M = C_T G \rho_{Mega} \qquad (9.8)$$

Robertson-Walker volume:

$$V = 2\pi^2 R^3 \qquad (10.1.2b)$$

$$\alpha = \gamma/\delta = 1 \qquad (10.6.3)$$

From the current CMB temperature ($T_0 = 2.7255\ ^\circ K$) we find

$$\kappa T = \kappa T_0/a(t_{now}) = \kappa T_0 = 2.35 \times 10^{-4}\ ev \qquad (13.3.1.1)$$
$$\chi = \pi^{3/2}\kappa T_0/(2M_c) \cong 5.3 \times 10^{-32} \qquad (13.3.1.2)$$

$$H_0 = 100\ h\ km\ s^{-1}\ Mpc^{-1} = h \times (9.777752\ Gyr)^{-1} =$$
$$= 1.1 \times 10^{-28}h\ cm^{-1} \equiv 3.24 \times 10^{-20}h\ s^{-1} \qquad (13.2.1.2b)$$
$$= 5.56 \times 10^{-57}\ cm^{-2} \qquad (13.2.1.2c)$$

$$\Omega_m + \Omega_\Lambda \cong 1.00$$

$$\xi = k/H_0^2 \cong 1 \qquad (13.2.1.3)$$

$$a(t_{now}) = 1 \qquad (13.2.2.4)$$

$$H_0\Omega_\Lambda^{1/2} = 1.83 \times 10^{-18}\ s^{-1} \qquad (13.2.2.6)$$

$$t_E = t_{now} + (H_0\Omega_\Lambda^{1/2})^{-1}\ln(.76) = 2.87 \times 10^{17}\ s \qquad (13.2.2.7)$$
$$t_{now} = 4.35 \times 10^{17}\ s$$
$$t_{RM} = 7 \times 10^{11}\ s = 2.22 \times 10^{-5}\ Gyr \qquad (13.2.3.12)$$
$$t_c = 1.26 \times 10^{-165}\ s \cong 10^{-165}\ s \qquad (13.3.2.15)$$
$$t_T = 380,000\ years$$
$$= 1.19837 \times 10^{13}\ sec$$
$$= 1.205 \times 10^{-11}\ Gyr$$

$$\kappa T = \kappa T_0/a(t_{now}) = \kappa T_0 = 2.35 \times 10^{-4}\ ev \qquad (13.3.1.1)$$

$$\varpi = k^{-1/2}M_c = 8.31 \times 10^{60} \qquad (13.3.2.2)$$

$$\gamma = \chi/\varpi = 6.38 \times 10^{-93} \qquad (13.3.2.9)$$

Appendix C. Evidence for Entities Beyond Our Universe

C.1 Theoretical Evidence for Other Universes and External Matter

Why are we not content with one universe given its enormous size and variety? It appears that there are important theoretical reasons, and some important experimental observations, that suggest that there is more than our universe 'out there.' The external entities are likely to be other universes but external "clumps" of mass-energy are not excluded.

In this appendix[29] we will discuss theoretical reasons and experimental suggestions of a larger space—that we call the *Megaverse*—that contains our universe and, most likely, other universes. The existence of a Megaverse resolves important theoretical issues and may resolve some important astronomical puzzles that have appeared in recent years.

The theoretical issues, which have been subjects of discussion for some time, are:

1. The need for an external 'clock' to measure 'time' knowing that it is to some extent relative and local.
2. The need for a 'quantum observer' to complete the understanding of quantum gravity as described by the Wheeler-DeWitt equation and other efforts to develop a quantum gravity.
3. The need for other universes to provide theoretical measuring platforms for quantities beyond the charge and mass-energy of the universe. We think here of the other quantum numbers of particles and particle number operators such as Baryon number.

---

[29] Most of this chapter appears in Blaha (2015a) and in earlier books by the author.

4. The need for an ultimate source of mass and inertia in our universe.

In Blaha (2015a), and earlier books, we have suggested that there are weighty reasons to believe that other universes exist.[30] The existence of other universes solves these problems.

These problems generally have a source in Quantum Gravity and the interpretation of the Wheeler-DeWitt equation in particular. See Blaha (2017c) and (2018e) for discussions of the Wheeler-DeWitt equation and its implications. We now consider the above issues.

C.1.1 Universe Clocks

Asynchronous Logic provides the equivalent of a clock for the synchronization of processes within large electrical systems such as VLSI chips. Similarly there is a need for a universal clock for our universe. As DeWitt[31] points out in his studies of quantum gravity,

'"The variables … [of the quantized Friedmann model] because of their lack of hermiticity, are not rigorously observable and hence cannot yield a measure of proper time which is valid under all circumstances. … . It is for this reason that we may say that "time" is only a phenomenological concept … If the principle of general covariance is truly valid then the quantum mechanics of everyday usage with its dependence on the Schrödinger equations … is only a phenomenological theory. For the only "time" which a covariant theory can admit is an intrinsic time defined by the contents of the universe itself. Any intrinsically defined time is necessarily non-Hermitean, which is equivalent to saying that there exists no clock, whether geometrical or material, which can yield a measure of time which is operationally valid under *all* circumstances, and hence there exists no operational method for determining the Schrödinger state function with arbitrarily high precision."

---

[30] In Blaha (2013a), before the Higgs particle was discovered at CERN we suggested an alternate mechanism was possible if a sister universe existed (making the existence of other universes a reasonable possibility. The Higgs discovery makes the sister universe mechanism unlikely.

[31] DeWitt, B. S., Phys. Rev. **160**, 1113 (1987).

The lack of a clock within our universe invalidates quantum mechanics in principle and Quantum Gravity in particular. DeWitt concludes, "Thus [quantum gravity] will say nothing about time unless a clock to measure time is provided." And it appears absolutely necessary.

Unruh[32] also has an issue with the source of time:

"One of the key problems is that of time. We see and experience the world in terms of time. We see things grow, develop, and change. However, time does not enter into the Euclidean formulation of quantum gravity directly. In the usual Hamiltonian formulation, the Hamiltonian for quantum gravity is made up of densities which are the generators, not only of spatial coordinate transformations, but also of temporal coordinate transformations. The content of four of Einstein's equations is that some generators are zero. Thus all wave functions are invariant under all spatial and all temporal coordinate transformations. There is nothing in the wave function or the amplitudes which refers to the coordinate t, or the corresponding points of the manifold in any way. How then do we recover the indubitable and ubiquitous experience we have of time? The standard answer is that our experience of time is actually an experience of different correlations between physical quantities in the world. Time is replaced by the readings of clocks. I know that time has changed, not through any direct experience with time, but because the hands of my watch have changed.

Although the implementation of this idea is actually extremely difficult in practice, and although I personally believe that one should formulate one's quantum theory of gravity so as to contain time explicitly, let us nevertheless pursue the consequences of this idea of time as defined internally, as the "reading" of a dynamic variable. For an observer inside the theory, his "time" is not the coordinate; rather his time is some one of the given dynamic variables of the theory: y or P. Thus although the coupling to the baby universes via the effective action S is independent of the coordinates t or x, that does not mean that the observer inside the theory will experience the interactions as being independent of time. For him and/or her, time is one of the dynamic variables and so it can depend on the various dynamic variables of the theory,

---

[32] Unruh, W. G., Phys. Rev. D **40**, 1053 (1989).

even if it does not depend on the time coordinate t. In general one would expect the observer to see what looks to him like a time-dependent interaction with the baby universes. At one time, some one of the baby universes may couple strongly to the large universe, while at some other time, another of the baby universes will couple more strongly."

In Blaha (2015a) and earlier books, we suggested the existence of other universes provides a 'clock mechanism' in principle for our universe. And being universes themselves, these other universes are excellent clocks. DeWitt points out,

"Because every clock has a "one-sided" energy spectrum, its ultimate accuracy must necessarily be inversely proportional to its rest mass. When the whole universe is cast in the role of a clock, the concept of time can of course be made fantastically accurate (at least in principle) ... "

Setting a mass scale using other universes, also sets[33] a time scale and resolves the issue of a clock for our universe. *In principle the existence of other universes validates the role of time in the Copenhagen interpretation of Quantum Mechanics.*

C.1.2 Quantum Observer

Attempts to create a quantum gravity theory have to confront the need for an *Observer* in any quantum theory within the context of the Copenhagen interpretation. DeWitt points out,

"The Copenhagen view depends on the assumed a priori existence of a classical level to which all questions of observation may ultimately be referred. Here, however, the whole universe is the object of inspection; there is no classical vantage point, and hence the interpretation question must be re-argued from the beginning. While we do not wish to stress this point unduly, since, after all, the Friedmann model ignores the

---

[33] For example the Planck time value is set by the Planck mass.

vast complexities of the real universe, it is nevertheless clear that the quantum theory of space-time must ultimately force a deviation from the traditional Copenhagen doctrine."

And Unruh states

"One of the key features in the interpretation of such transition amplitudes, or wave functions, is the idea that we, as observers are also a part of the Universe as a whole. We, as physical observers, must be describable from within the theory and not as observers external to the theory as in usual quantum mechanics. In usual quantum mechanics, the interpretation is usually given in terms of observers that are outside of the theory. There one makes a split, with the quantum world at one side of the split, and the observer on the other. von Neumann argued that the predictions of quantum mechanics, at least under certain assumptions, are independent of the exact location of that split, but Bohr argued adamantly for the necessity of such a split (classical observers and quantum world). *There is a great difficulty in setting up such a split for physical observers contained within and influenced by a quantum universe,* [italics added] and for the Universe as a whole, especially including gravity, one cannot argue that the predictions will be independent of where one puts the split. Since all energies interact gravitationally, and our observations are surely energetic phenomenon, the treatment of the energetics of observation as classical would lead to different predictions than if they were treated quantum mechanically. One is therefore forced to devise an interpretation of quantum mechanics in which the observer is part of the quantum system, rather than outside the quantum system.

This means that the interpretation of these transition amplitudes becomes somewhat non-intuitive. One must ask what the system looks like from within, from the viewpoint of an observer who is part of that world, rather than being able to interpret them directly in terms of probabilities for observations made by an external observer."

While the *Observer* question is addressed by a number of authors, the proposed answers are not entirely convincing. *The existence of other universes provides macroscopic Quantum Observers for our universe.* And our universe acts as a macroscopic quantum observer for other universes. Thus the quantum observer issue is resolved within the Megaverse of universes.

These considerations lead us to view the existence of other universes as a critical solution to the above problems.

C.1.3 The Higgs Mechanism is Explainable by Extra Dimensions

The Higgs Mechanism 'explains' (generates) fermion and boson masses. However the Higgs potential contains a quadratic term with a constant with the dimensions of [mass]. In a sense the Higgs Mechanism trades one mass for another. From where do the Higgs potentials' masses come?

A further explanation is needed is to determine the origin of the "dimensionful" mass terms in the Higgs' particle equations themselves. At present little if any thought has been given to the origin of these terms. We suggested that, excluding a *deus ex machina* source, the only known way to generate these mass terms in the Higgs' equations is through the separation of equations technique of differential equations. This technique requires additional parameters which can only be the coordinates of *extra unknown dimensions*. The best example of the generation of mass terms appears in the Schwarzschild solution of General Relativity where a separation constant, often denoted M, appears that has the dimension of [mass].

Thus extra space-time dimensions would resolve the origin of Higgs potentials' masses. Given extra dimensions it is reasonable to expect that these extra dimensions contain universes. Thus the Megaverse!

C.1.4 Possible Accretion of Megaverse Matter to Fuel Expansion of Our Universe

If matter is distributed outside of universes in the Megaverse, and if this matter can be accreted to universes by gravitational attraction, then the apparent increasing expansion of our universe may be due to this accretion. In chapter 14 of Blaha (2017c) we presented a model in which this possibility is realized. If true, then we would have tangible evidence of the existence of other universes in the Megaverse.

C.1.5 Asynchronous Logic is a Requirement of Universes

By establishing Asynchronous Logic principles[34] as the basis for the existence of universes and for setting the number of dimensions in each universe – four; and establishing the basis of fermion particles as *qubes*—we have found deeper principles of organization for the foundations of physics. The principles built on this foundation serve to enable the coordination of complex physical processes.

Usually we look at particle processes primarily from a space-time perspective: particles collide and produce new particles. We primarily think of the incoming and outgoing particles in a collision. However, considering the set of fundamental particles – and particle transforming interactions in themselves, while neglecting space-time and momentum considerations, leads us to view particles as constituting an alphabet and to view their interactions as a type of computer grammar.[35] Then the Asynchronicity Principles enable us to bring in space-time in a way that gives us the maximum complexity with the most minimal assumptions. As Leibniz[36] points out our universe has maximal complexity with minimal assumptions.

C.1.6 The Meaning of Total Quantities of a Universe

The 'external' properties of a universe are normally questioned—for the simple reason that it is assumed that there is no 'outside' of our universe. For example, Misner (1973) asserts:[37]

'There is no such thing as "the energy (or angular momentum, or charge) of a closed universe," according to general relativity, and this for a simple reason. To weigh something one needs a platform on which to stand to do the weighing."

---

[34] The basis of this section is described in detail in Blaha (2015a). That book places Physics within a logical framework that is a possible deeper ground for fundamental Physics theory.

[35] This conceptual approach was first described in Blaha (1998) who went on to characterize our universe as one enormous word evolving in time.

[36] See Rescher (1967).

[37] Pp. 457 - 458.

Misner et al presumes no such platform exists. If there is but one closed universe as most physicists currently believe then one cannot measure any totals of a closed universe such as ours. Yet if we take a more general view that our universe is only one of many then it becomes possible to measure total mass, charge, angular momentum, baryon number, and many other quantities of interest. Indeed, the existence of other universes (within the encompassing Megaverse) opens the door to an understanding of time, mass, energy, and all the other quantities necessary to develop a dynamical theory of universes.

Our new 'rotations of interactions' formalism (described in previous books) enables us to rotate measurable quantities. These quantities (quantum numbers) furnish a set of totals for our universe such as baryon numbers (normal and Dark), lepton numbers, angular momentum and so on that characterize our universe.

Later we will also see that one can then treat universes as 'particles', and develop 'universe dynamics', which might explain knotty problems such as the Big Bang and its precursor (if any). We will do this in subsequent chapters after first considering the possible structure of universes in general in the Megaverse.

C.2 Possible Experimental Evidence for the Megaverse

At first glance it would seem impossible to produce evidence for the existence of other universes. However there are subtle means by which we can 'sense' experimentally 'nearby' universes should they exist. The mechanism would appear to be gravitational effects exerted on objects within our universe by unseen nearby objects of enormous mass. Currently there appears to be three experimental suggestions for the existence of 'nearby' universes and one theoretical argument based on an influx of mass-energy from the Megaverse that may support an understanding of the expansion of our universe.

C.2.1 Great Attractors

One potential support is the discovery of the Great Attractor (at the center of the Laniakea Galaxy Supercluster), and the more massive Shapley Attractor (centered in the

Shapley Supercluster)[38]. These attractors contain massive numbers of galaxies and are drawing galaxies over a distance of millions of light years towards them.

If another universe(s) is 'near' our universe it could act as a 'gravitational magnet' and draw galaxies within our universe towards it to form one or more superclusters which could then act as attractors. Thus attractors might indirectly reveal the presence of other nearby universes—contrary to the expected large scale uniformity of the universe. The only other apparent source of superclusters is chance. Chance seems an unsatisfactory possibility in the present case.

C.2.2 Bright Bumps in Universe Suggesting Collision with Another Universe

A recent study[39] of the residual brightness of parts of the accessible universe found that bright patches appeared if a model of the CMB (Cosmic Microwave Background) with gases, stars and dust was 'subtracted' from the PLANCK map of the entire sky. After the subtraction one would expect only noise spread throughout the sky. However, bright patches were seen in a certain range of frequencies. These anomalies are thought to be a result of our universe colliding with another object – presumably another universe in the Megaverse.

C.2.3 Cold Spot in Universe Suggesting Collision with Another Universe

Another recent study[40] of a huge cold region of the universe spanning billions of light years revealed that this region is not a relatively empty region but rather is similar to in its distribution of galaxies to the rest of the universe. Previous the Cold Spot was considered an area where cosmic microwave background radiation – the leftover Big Bang radiation is weak – making it significantly colder (0.00015C colder) than the average temperature of the universe.

An analysis of 7,000 galaxy redshifts using new high-resolution data has now shown that the Cold Spot is similar to the rest of the universe. The Durham University

---

[38] Tully, R. Brent; Courtois, Helene; Hoffman, Yehuda; Pomarède, Daniel, "The Laniakea Supercluster of galaxies". Nature (4 September 2014). 513 (7516): 71–73; arXiv:1409.0880.

[39] Ranga-Ram Chary, arXiv.org:/1510.00126 (2015).

[40] T. Shanks et al, Durham University (Australia), Monthly Notices of the Royal Astronomical Society, 2016.

group suggested that the Cold Spot might have been caused by a collision between our universe and another Universe. They further suggested that there is only a 1 in 50 chance that it could explain this feature with standard cosmology.

Thus we have another important piece of circumstantial evidence in favor of other universes and thus the Megaverse.

C.2.4 Megaverse Energy-Matter Infusion into Our Universe

In Blaha (2019c), and earlier books, we presented a model for an influx of mass-energy from the Megaverse to support the Bondi-Gold-Hoyle-Narlikar Steady State Cosmology, which was originally based on the 'continuous creation of mass-energy' by Hoyle and Narlikar. This model explains why the value of Ω makes the universe close to flat. If this model is correct then we would have concrete support for a Megaverse with a low mass-energy density leaking mass-energy into our universe. *More generally, it suggests that universes are surfaces of high mass-energy density in a Megaverse of low mass-energy density – with a ratio of mass-energy densities of the other of 10^{30}.*

C.2.5 Conclusion

We conclude that data is beginning to emerge favoring multiple universes and a physical Megaverse in support of the theoretical justifications presented earlier.

C.3 Historical Trend Toward Larger Space-Time Structures

Looking back through the history of Mankind's view of the universe we see a clear progression to a larger and larger view. Before the 16th century the earth was the universe. In the 16th century Giordano Bruno (and possibly others) suggested that the stars were suns with many worlds circling them. So our view of the universe expanded to include stars. Now we are on the threshold of external universes.

C.4 Possibly Universe Fragments

In addition to other universes the possibility of diffuse mass-energy in the Megaverse appears likely. The possibility of "mini-universes" is also likely.

Appendix D. The Embedding of a Universe in a Higher Dimension Megaverse

D.1 Universes as Mass-Energy Islands

In developing the theory of the Megaverse we view universes as islands of mass-energy that maintain their 'integrity' as surfaces due to gravitation. Gravity holds universes together rather like molecular forces within a water droplet hold water molecules together. Molecular attractive forces give droplets cohesion as they (perhaps) descend through the earth's atmosphere. They are the origin of *surface tension* in water.

Similarly gravity holds higher density[41] mass-energy universes together and gives rise to gravity surface tension. In a model presented in Blaha (2017c) for the Big Bang and the expansion of the universe within the Megaverse we found the mass-energy density of the universe was a factor of 10^{30} more than the density in the surrounding Megaverse space.

Thus we have good reason to study the surface tension of universes as a result of gravitational attraction within universes.

D.2 Boundary of a Universe within the Megaverse

In this chapter we describe the embedding of universes within the Megaverse. Much of this chapter appears in several earlier books by the author such as Blaha (2015a).

As stated earlier, we define a universe to be a closed or open surface in the Megaverse with a much higher mass-energy density than the Megaverse.

There are two types of boundaries for a universe embedded in a space of larger dimensions. First there is a boundary of the universe determined by treating the universe

---

[41] Higher density in comparison to the much lower density of the inter-universe space of the Megaverse.

as a surface in the space. Secondly, there is another type of universe boundary defined by the observation that any neighborhood of ever

A point of a universe – not strictly within the universe – has an infinite number of points of the enclosing Megaverse space.[42] Or, every universe point has neighborhoods with Megaverse points within it. Thus *each point of a universe is on a boundary of the universe due to the larger dimensions of the Megaverse space* within which it resides. Fig. D.1 schematically illustrates these neighborhoods for any universe point for a universe contained within a higher dimensional Megaverse.

D.3 Confinement of Universes due to 'Surface Tension'

We will assume that other universes have the same physics as our universe with the possible differences that they may have differing interaction coupling constants and differing particle masses. As we stated above, every point of a universe in a higher dimensional Megaverse has Megaverse points in any neighborhood of the point (with the exception of neighborhoods strictly within the universe). Thus we confront the question: what keeps mass-energy at points in a universe, or is there leakage from the universe into the Megaverse?

If there is little or no leakage into the Megaverse then, since each point in a universe is part of a Megaverse surface, one can only assume that there is a barrier to movement into the Megaverse. Taking a note from fluid dynamics, and viewing Megaverse space as one 'material' and the universe as a different 'material,'[43] we view the barrier as 'surface tension.'[44] The Megaverse appears to "exert a force" confining the contents of the universe to within itself.[45]

---

[42] Any neighborhood of any point in the universe – with all its points strictly within the universe – has an infinite number of points within the universe. We assume the neighborhood is so small that the curvature of the universe's space can be neglected.
[43] Meaning material with much higher mass-energy density and consequently larger internal gravitational attraction.
[44] See Landau (1987).
[45] Although in actuality it is the universe that holds itself together by gravitation. The surface tension force is the result of the internal gravitation of mass-energy within the universe.

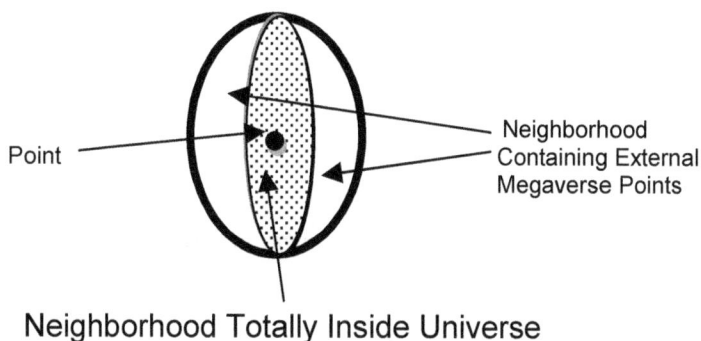

Neighborhood Totally Inside Universe

Figure D.1. Schematic diagram of a 3-dimensional projection of 'orthogonal' neighborhoods of a point within a universe with one neighborhood strictly within the universe and the other neighborhood containing external Megaverse points in general. The 'orthogonal' circles around the point differentiate between the two types of neighborhoods.

The surface tension[46] of a universe γ satisfies the relation

$$\gamma = W/\Delta A \qquad (D.1)$$

where γ is expressed in erg/cm^2, W is the Work, and ΔA is the Area upon which the work is exerted. The pressure Δp exerted by the surface tension for a 'spherical' surface area is

$$\Delta p = 2\gamma/R \qquad (D.2)$$

where R equals the radius of curvature of the surface. The above equations embody the concept that the surface tension force equals the pressure difference at the surface.

---

[46] A useful analogy: the Megaverse is a pool of water; a universe is a denser oil bubble within it. Surface tension caused by the cohesiveness of the oil molecules in the bubble makes it spherical (confines it to a spherical shape). Similarly a universe (denser than the Megaverse) is 'confined' within the Megaverse.

If the universe is flat then the surface pressure approaches ∞ giving confinement of the fields and particles to the universe:

$$R \rightarrow 0 \quad \text{implies } \Delta p \rightarrow \infty \tag{D.3}$$

Thus we have the theorem:

Theorem: A universe has no leakage of fields or particles into a higher dimensional space if the universe is exactly flat.

This theorem is particularly interesting in the case of our universe. It appears to be flat (or very close to flat). The flatness of our universe may be the reason no leakage of fields or particles from our universe has been detected to high accuracy.

If a universe is found with a non-zero radius of curvature then one can expect that some fields and particles may emerge from it into the Megaverse.

While a zero radius of curvature prevents the exit of fields and particles from a universe, it does not prevent the entry of mass-energy into the universe from the Megaverse. Thus our Continuous Creation Model of chapter 14 in Blaha (2017c) may be relevant. Entry is possible; exit is forbidden in this case.

Eqs. D.1 – D.2 are reminiscent of the 'four laws for Black Holes' which are stated later.

D.4 Quantum Fields 'Emanating' from a Universe

If the curvature of the universe is zero (open universe) then no fields emanate from it. If the curvature of the universe is non-zero (closed universe) then fields may 'leak' into the Megaverse. Then continuity conditions between a universe field and its Megaverse counterpart becomes of interest. We discuss this in detail later.

D.5 Universe Confinement by Conservation Laws

Every point in our universe is "infinitely" close to points of the Megaverse.

A universe occupies a region within the Megaverse. However because it is a lower dimension surface within the Megaverse the neighborhood of every point within a universe has an infinite number of Megaverse points that are not within the universe.

One might think that particles within a universe could then 'slip' into the Megaverse outside the universe with ease. However that is not the case. The law of momentum conservation compels particles and interactions within a universe to be confined to the universe. More importantly, the Megaverse surface tension of a flat universe confines particles and fields within a universe.[47]

The only possible ways that a particle could exit from a universe are 1) if the particle collides with a particle with a momentum, some of whose components are in Megaverse dimensions extraneous to the universe's dimensions; or 2) a particle within the universe experiences forces with components in Megaverse dimensions extraneous to the universe. We shall consider the second possibility later when we consider a mechanism for a starship to exit our universe (chapter 5). The first possibility exists if the Megaverse has a very low matter density outside of universes. The fact that this phenomenon has not been observed implies the Megaverse matter density is extremely low.

Thus conservation of momentum for particles and interactions, and surface tension, effectively confines particles within a universe even though the neighborhood of every point of a particle's trajectory contains an infinity of Megaverse points exterior to the universe. Similarly every interaction within a universe is confined when expanded in a Fourier series (assuming free fields) in universe coordinates.

It is also possible for a Megaverse particle to enter[48] a universe through perhaps a collision that results in the particle being within the universe with momentum also solely within the universe. Thus the 'point boundary' of universes is porous. Particles can enter/exit a universe under appropriate conditions.

---

[47] A useful analogy: the Megaverse is a pool of water; a universe is a denser, oil bubble within it. Surface tension caused by the cohesiveness of the oil molecules in the bubble makes it spherical (confines it to a spherical shape).

[48] In chapter 14 of Blaha (2017c) we consider a model that supplies a mechanism for the Hoyle-Narlikar continuous creation theory for the expansion of our universe. This model 'creates' mass-energy as an inflow from the Megaverse.

We conclude the 'point boundary' of a universe is not a barrier although surface tension force controls the entry/exit of particles and fields. We consider an exit mechanism from a universe in chapter 5.

D.6 Objects Straddling a Universe-Megaverse Boundary

When a starship, or some other extended object, is entering/exiting a universe at some velocity the question of the state of the object arises. It is partially in and partially out of the universe. We know that the object being 4-dimensional will continue to be 4-dimensional, barring effects of forces that might "twist" parts of the object into additional dimensions.

There are also the more subtle quantum effects on the object due to the possibility of different quantization of the particles in a universe and the Megaverse. Quantization in different coordinate systems might result in different physical interpretations of matter. We resolve this issue by our PseudoQuantization Method described in Blaha (2018e). In its Appendix A we show that one can quantize using a form of Pseudoquantization that preserves (unitarily equivalent) particle interpretations in a universe and the Megaverse.

Thus extended objects can be partly in a universe, and partly in the external Megaverse without issues.

Appendix E. General Properties of the Megaverse

If one wishes to have a portrait of the Megaverse it seems likely that the it would contain universes that were scattered within it in a fashion similar to the scattering of galaxies within our universe—although on a much larger distance scale in multiple dimensions. In this chapter we will overview properties of the Megaverse and its universes. Much of this material previously appeared in Blaha (2017c) and in earlier Physics and starship travel books.

E.1 Megaverse Size, Lifetime, and Universe Separation

The first questions that naturally arise are the age, size, and general characteristics of universes within the Megaverse. In the absence of any experimental detail we will assume the relative size of entities (universes) in the Megaverse[49] is proportionate to the relative size of the entities (galaxies) in the universe.

(Average Galaxy Size)/(Universe Size) = (Average Universe Size)/(Megaverse Size)   (E.1)

Taking the average diameter of galaxies to be 400,000 light years, the age of the universe to be 13,800,000,000 light years, and the diameter of the universe to be 91.4 billion light years[50] (the estimated diameter of last scattering surface) we find the diameter of the Megaverse to be

$$\text{Diameter}_{\text{Megaverse}} = 2 \times 10^{16} \text{ light years} \qquad (E.2)$$
$$= 228,500 \times \text{Diameter}_{\text{Universe}}$$

---

[49] We assume the Megaverse is homogeneous at large distance scales. We also assume a proportionality between the average entity size (of galaxies in the case of unbiverses; and of umiverses in the case of the Megaverse) in part based on the relatively long lifetimes of universes and the Megaverse.

[50] There are much larger estimates of the universe's diameter based on the Inflation theory of A. Guth and others.

Using the 228,500 scale factor, we find the Megaverse age since the Megaverse purported 'Big Bang' to be

$$Age_{Megaverse} = 3 \times 10^{15} \text{ years} \qquad (E.3)$$
$$= 3 \text{ million billion years}$$

If the average separation between galaxies in our universe is 3,000,000 light years, then assuming distance scaling by 228,500, the average separation between universes in the Megaverse would be

$$Separation_{Universes} = 228,500 \times 3,000,000 \text{ ly} = 7 \times 10^{11} \text{ light years} \qquad (E.4)$$
$$= 700 \text{ billion light years}$$

If we now assume the mass of a universe equals the total mass-energy of our universe[51] (including Dark mass and Dark energy) which is estimated to be $m_{universe} = 3 \times 10^{54}$ kg then the gravitational potential energy between two such universes separated by 700 billion light years is

$$V = G \; m_{universe}^{2}/Separation_{Universes} \qquad (E.5)$$
$$= 9 \times 10^{70} \text{ kgm}^2\text{s}^{-2}$$

Then the gravitational force is

$$F = G \; m_{universe}^{2}/Separation_{Universes}^{2} \qquad (E.6)$$
$$= 1.35 \times 10^{43} \text{ kgms}^{-2}$$

and the resulting gravitational acceleration of universes is

$$a = E.5 \times 10^{-12} \text{ m/s}^2 \qquad (E.7)$$

The Baryonic and Leptonic forces associated with the Generation group of the Unified SuperStandard Theory may slightly modify the force between the universes. We view

---

[51] We assume our universe is 'average' in mass and size for the sake of discussion.

the small acceleration between universes due to gravity as physically acceptable. In a billion years the universe velocity would be $v = 1.4 \times 0^5$ m/s with v/c = 0.0005 and the distance traveled equal to 2.2×10^{21} m = 236,540 light years – negligible compared to the $Separation_{Universes}$. Universes would, on the average, only make contact after extraordinary long times. Thus we have a preliminary coherent view of Megaverse size and distance parameters.

E.2 Likely Features of the Megaverse

There are a number of features of the Megaverse that appear to be true.

E.2.1 Megaverse Curvature

The Megaverse has gravitation. Gravitation appears to be weak in the Megaverse so it is close to a flat space. The sources of Megaverse gravitation are the mass-energy of the universes within it and the density of mass-energy of particles outside of universes. Universes have a larger relative force of gravity due to a higher mass-energy density. Universes therefore have more curvature.

E.2.2 Megaverse Time Dimension

We will assume that the Megaverse has one complex time dimension denoted y^D for the simple reason that the absence of a time dimension would make the Megaverse static.

E.2.3 Megaverse Forces

In addition to Megaverse gravitation, the Megaverse has the forces in the Unified SuperStandard Model. These forces satisfy continuity conditions at universe boundaries.

E.2.4 Megaverse Parameters

Megaverse physical constants and particle masses have the same values as in our universe due to continuity.

E.2.5 Megaverse Vacuum Fluctuations

Megaverse Vacuum fluctuations may be a source of the generation of universes and particles. Vacuum fluctuations might account for the Big Bang. The time scale for the persistence of universes generated by a vacuum fluctuation is likely to be an extrapolation of vacuum fluctuation persistence within our universe.

E.2.6 Megaverse Matter and Chemistry

The existence of many more dimensions in the Megaverse suggests that multi-dimensional forms of matter and energy could exist *between* universes. As a result such Megaverse atoms, compounds and Chemistry will be very different and much more varied than in our universe. If such matter exists in the Megaverse then 'mining' such matter for use in our universe—would give us exotic new compounds and Chemistry that would be partially inside and partially outside, of our universe.

This possibility makes venturing into the Megaverse economically and scientifically desirable since such materials cannot be created within our universe.

E.3 Features of Universes within the Megaverse

We know of our universe from the 'inside.' However the features of our universe from a Megaverse perspective are not at all certain. In this section we will describe the Megaverse view of a universe's properties.

We shall assume a universe is a closed or open surface within the Megaverse of much higher mass-energy density than the Megaverse's mass-energy density by perhaps as much as a factor of 10^{30}. Chapter 14 of Blaha (2017c) describes a model of Megaverse mass-energy inflow into our universe exemplifying this feature.

E.3.1 Universe Area and Mass

The mass of a universe is an important property since mass is one of the sources of Megaverse gravitation, and interaction between universes.

While a universe is not believed to be a black hole (although Hawking has recently jokingly? suggested that our universe may be a black hole, and even more recently suggested black holes are not quite black holes – grey?), there are general qualitative similarities that lead us to consider the possibility that the four laws of black holes[52] may apply in part (or their entirety) to universes. In particular the 2nd law states

$$dM = \kappa dA/8\pi + \Omega dJ \qquad (E.8)$$

where dM is the change in "mass/energy," A is the area of the Black Hole (universe), Ω is its angular velocity and J is the angular momentum.[53] From eq. E.8 it appears we can reasonably define a "mass" for a universe in terms of a universe's area:

$$M = \kappa A/8\pi \qquad (E.9)$$

This definition seems to capture the physics of universes that could be used in developing a dynamics of universes as we do later. It allows us to escape the dilemma of having zero total energy for universes that would preclude treating universes as particles in the Megaverse and developing a Megaverse dynamics of universe-particles. Later we will also show how to define a mass for a universe that is time dependent.

E.3.2 Relation between Universe and Megaverse Vector Fields

In Blaha (2018e) we defined the interactions of The Unified SuperStandard Theory. These interactions and their groups also exist in the Megaverse. The Fourier expansions of fields in the Megaverse are different. They must be expressed in terms of the D coordinates of the Megaverse.

Within a universe, since each point of the universe is surrounded by Megaverse points, a field has both an expression in universe coordinates, x, and an expression in Megaverse coordinates, denoted y. For a vector field in the universe $A_U^{\mu}(x)$ there is an equivalent representation of the field in Megaverse coordinates $A_M^{i}(y)$. These representations are related by a coordinate transformation. If we define a map from universe coordinates to Megaverse coordinates with

---

[52] Wald, R. M., "The Thermodynamics of Black Holes", *Living Reviews in Relativity* **4** (6): 12119 (2001).

[53] Although the angular momentum of a universe is not measurable if there is only one universe (as DeWitt argued in a quote earlier), the existence of multiple universes within the Megaverse enables the relative angular momentum of a universe to be determined.

$$y^i = f^i(x) \tag{E.10}$$

then the field representations are related by[54]

$$A_M^i(y) = \partial y^i / \partial x^\mu \, A_U^{\;\mu}(x) \tag{E.11}$$
$$= \partial y^i / \partial x^\mu \, A_U^{\;\mu}(f^{-1}(y))$$

in the domain of the universe. The values of the field at Megaverse points in a neighborhood of a universe point are determined by continuity.

Outside the domain of a universe the Megaverse field value is determined by its sources.

At the boundary of a universe there must be continuity in the expectation values of the Megaverse fields.

E.3.3 Megaverse Gravitation and Free Matter

We assume that a generalization of Einstein's theory of Gravity exists in the Megaverse.

Just as our universe has matter and radiation between galaxies, it seems reasonable to assume that 'free' matter and radiation exists in the Megaverse outside of universes. Such mass-energy would have two roles: to gravitationally affect the dynamics of the Megaverse and the motion of universes within it, and to possibly fuel the expansion of universes. The expansion of our universe may be due to an influx of matter and energy from the external Megaverse. Many years ago Hoyle and Narlikar considered the possibility of 'continuous creation of matter.' We suggest that an influx of Megaverse matter may be the actual source. We considered this possibility in chapter 14 of Blaha (2017c), which contains an unpublished paper by this author written approximately seven years previously.

---

[54] Implicit in eq. 4.10 is an inverse relation $x^\mu = f^{-1}(y)$ which is necessarily based on a restriction of the y-coordinates to obtain a 1:1 relation between the y and x coordinates. The restriction is best implemented by requiring the domain of y coordinates be restricted to those y coordinates within the universe surface. The result is a 1:1 relation between the y-domain coordinates and the x universe coordinates.

Thus we arrive at a view of the Megaverse of matter and universes that is analogous to our universe of galaxies.

E.3.4 Possible Source of Expansion of Universes

Our universe expanded from the Big Bang to its current size and is still expanding. It is likely that other universes have undergone similar expansions. According to chapter 32 of Blaha (2018e) there is an infinite surface tension at the boundary of our universe. How can the universe have expanded, and continue expanding, under such conditions. We see a two phase expansion of the universe.

For a period of time after the Big Bang the universe did not have an infinite surface tension, thus hindering universe expansion into the Megaverse due to surface tension force. The cause is an effect discovered by Eőtvos – a temperature dependence of the surface tension force. Eőtvos pointed out a critical temperature T_c existed that caused the surface tension force to decline as the temperature increased:

$$\gamma V^{2/3} = k(T_c - T) \tag{E.12}$$

where k is the Eőtvos constant, and V is the volume of the universe (the liquid 'drop') Assuming a spherical universe, the volume is

$$V = 4\pi R^3/3. \tag{E.13}$$

Thus

$$\gamma = (4\pi/3)^{-2/3} k(T_c - T)R^{-2} \tag{E.14}$$

For very high temperatures such as existed after the Big Bang $T > T_c$ γ would be negative indicating that an outward pressure existed from the universe into the Megaverse promoting expansion. Thus in the high temperature period ($T > T_c$) after the Big Bang the surface tension force favored expansion of the universe.

After this phase, the surface tension γ is positive. The Megaverse is then superficially partially 'impeding' expansion due to surface tension. However, the surface tension pressure of the Megaverse causes leakage *into* the universe from the

Megaverse causing its mass to increase, and its radius and volume to increase. Expansion due to the accretion of Megaverse mass-energy!

The above scenario is supported by the two phase hybrid model consisting of a Big Bang expansion model, and a mass-energy accretion model – all of Blaha (2019c). Note as the radius of curvature goes to zero, eq. E.14 suggests an increasing surface tension pressure 'pushing' particles into the universe.

Thus a complete universe expansion scenario is evident based on surface tension physics and the Big Bang.

E.3.5 Universe Generation from Vacuum Fluctuations

Vacuum fluctuations could generate universe-antiuniverse pairs. Antiuniverses would have certain 'negative quantum numbers.' We view universes/antiuniverses, when created (Big Bangs), as 'ultra-small' particles of large mass-energy that can proceed to expand to great size. If they are generated by a vacuum fluctuation they will, after possibly a certain time, recombine into the vacuum.

E.3.6 Universes as Black Holes

It is conceivable that a universe could be so dense and confined that it would be effectively a Black Hole. In this case the internal quantum numbers of the Black Hole universe would be inaccessible and only it's mass, velocity and angular momentum would be observables.

E.3.7 Life in Other Universes

It is likely that life exists in some if not all of the universes of the Megaverse. If the physical constants, laws, and masses of the interior of a universe are similar if not identical to ours then the possibility of intelligent species, even human-like species, is very likely. This possibility is an important motivation for humanity to reach for the Megaverse.

REFERENCES

Akhiezer, N. I., Frink, A. H. (tr), 1962, *The Calculus of Variations* (Blaisdell Publishing, New York, 1962).

Bjorken, J. D., Drell, S. D., 1964, *Relativistic Quantum Mechanics* (McGraw-Hill, New York, 1965).

Bjorken, J. D., Drell, S. D., 1965, *Relativistic Quantum Fields* (McGraw-Hill, New York, 1965).

Blaha, S., 1998, *Cosmos and Consciousness* (Pingree-Hill Publishing, Auburn, NH, 1998).

_____, 2018e, *Unification of God Theory and Unified SuperStandard Model THIRD EDITION* (Pingree Hill Publishing, Auburn, NH, 2018).

_____, 2019a, *Calculation of: QED α = 1/137, and Other Coupling Constants of the Unified SuperStandard Theory* (Pingree Hill Publishing, Auburn, NH, 2019).

_____, 2019b, *Coupling Constants of the Unified SuperStandard Theory SECOND EDITION: We Find the Fine Structure Constant 1/137.0359801, and so: OUR UNIVERSE AND LIFE! Also a Universal Eigenvalue Function for all Known Interactions, And Running Coupling Constants to all Perturbative Orders* (Pingree Hill Publishing, Auburn, NH, 2019).

_____, 2019c, *New Hybrid Quantum Big_Bang-Megaverse_Driven Universe with a Finite Big Bang and an Increasing Hubble Constant* (Pingree Hill Publishing, Auburn, NH, 2019).

Gelfand, I. M., Fomin, S. V., Silverman, R. A. (tr), 2000, *Calculus of Variations* (Dover Publications, Mineola, NY, 2000).

Gradshteyn, I. S. and Ryzhik, I. M., 1965, *Table of Integrals, Series, and Products* (Academic Press, New York, 1965).

Heitler, W., 1954, *The Quantum Theory of Radiation* (Claendon Press, Oxford, UK, 1954).

Huang, Kerson, 1992, *Quarks, Leptons & Gauge Fields 2nd Edition* (World Scientific Publishing Company, Singapore, 1992).

Kaku, Michio, 1993, *Quantum Field Theory*, (Oxford University Press, New York, 1993).

Landau, L. D. and Lifshitz, E. M., 1987, *Fluid Mechanics 2nd Edition*, (Pergamon Press, Elmsford, NY, 1987).

Misner, C. W., Thorne, K. S., and Wheeler, J. A., 1973, *Gravitation* (W. H. Freeman, New York, 1973).

Streater, R. F. and Wightman, A. S., 2000, *PCT, Spin, Statistics, and All That* (Princeton University Press, Princeton, NJ 2000).

Weinberg, S., 1972, *Gravitation and Cosmology* (John Wiley and Sons, New York, 1972).

Weinberg, S., 1995, *The Quantum Theory of Fields Volume I* (Cambridge University Press, New York, 1995).

Weinberg, S., 2000, *The Quantum Theory of Fields Volume III Supersymmetry* (Cambridge University Press, New York, 2000).

Weyl, H., 1950, *Space, Time, Matter* (Dover, New York, 1950).

INDEX

About the Author

Stephen Blaha is a well-known Physicist and Man of Letters with interests in Science, Society and civilization, the Arts, and Technology. He had an Alfred P. Sloan Foundation scholarship in college. He received his Ph.D. in Physics from Rockefeller University. He has served on the faculties of several major universities. He was also a Member of the Technical Staff at Bell Laboratories, a manager at the Boston Globe Newspaper, a Director at Wang Laboratories, and President of Blaha Software Inc. and of Janus Associates Inc. (NH).

Among other achievements he was a co-discoverer of the "r potential" for heavy quark binding developing the first (and still the only demonstrable) non-abelian gauge theory with an "r" potential; first suggested the existence of topological structures in superfluid He-3; first proposed Yang-Mills theories would appear in condensed matter phenomena with non-scalar order parameters; first developed a grammar-based formalism for quantum computers and applied it to elementary particle theories; first developed a new form of quantum field theory without divergences (thus solving a major 60 year old problem that enabled a unified theory of the Standard Model and Quantum Gravity without divergences to be developed); first developed a formulation of complex General Relativity based on analytic continuation from real space-time; first developed a generalized non-homogeneous Robertson-Walker metric that enabled a quantum theory of the Big Bang to be developed without singularities at t = 0; first generalized Cauchy's theorem and Gauss' theorem to complex, curved multi-dimensional spaces; received Honorable Mention in the Gravity Research Foundation Essay Competition in 1978; first developed a physically acceptable theory of faster-than-light particles; first derived a composition of extrema method in the Calculus of Variations; first quantitatively suggested that inflationary periods in the history of the universe were not needed; first proved Gödel's Theorem implies Nature must be quantum; provided a new alternative to the Higgs Mechanism, and Higgs particles, to generate masses; first showed how to resolve logical paradoxes including Gödel's Undecidability Theorem by developing Operator Logic and Quantum Operator Logic; first developed a quantitative harmonic oscillator-like model of the life cycle, and interactions, of civilizations; first showed how equations describing superorganisms also apply to civilizations. A recent book shows his theory applies successfully to the past 14 years of history and to *new* archaeological data on Andean and Mayan civilizations as well as Early Anatolian and Egyptian civilizations.

He first developed an axiomatic derivation of the form of The Standard Model from geometry – space-time properties – The Unified SuperStandard Model. It unifies all the known forces of Nature. It also has a Dark Matter sector that includes a Dark ElectroWeak sector with Dark doublets and Dark gauge interactions. It uses quantum coordinates to remove infinities that crop up in most interacting quantum field theories and additionally to remove the infinities that appear in the Big Bang and generate inflationary growth of the universe. It shows gravity has a MOND-like form without sacrificing Newton's Laws. It relates the interactions of the MOND-like sector of gravity with the r-potential of Quark Confinement. The axioms of the theory lead to the question of their origin. We suggest in the preceding edition of this book it can be attributed to an entity with God-like properties. We explore these properties in "God Theory" and show they predict that the Cosmos exists forever although individual universes (or incarnations of our universe) "come and go." Several other important results emerge from God Theory such a functionally triune God. The Unified SuperStandard Theory has many other

important parts described in the Current Edition of *The Unified SuperStandard Theory* and expanded in subsequent volumes.

In 2019 Blaha calculated the correct value of the QED Fine Structure Constant based on his 1973 vacuum polarization paper. He also calculated approximate values for Standard Model U(1), SU(2), and SU(3) interactions, He also extended his 2004 model of the quantum Big Bang to the entire expansion of the universe using a new universal scale factor formulation. This formulation appears to be analogous to QED vacuum polarization..

Blaha has had a major impact on a succession of elementary particle theories: his Ph.D. thesis (1970), and papers, showed that quantum field theory calculations to all orders in ladder approximations could not give scaling deep inelastic electron-nucleon scattering. He later showed the eigenvalue equation for the fine structure constant α in Johnson-Baker-Willey QED had a zero at $\alpha = 1$ not $1/137$ by solving the Schwinger-Dyson equations to all orders in an approximation that agreed with exact results to 4th order in α thus ending interest in this theory. In 1979 at Prof. Ken Johnson's (MIT) suggestion he calculated the proton-neutron mass difference in the MIT bag model and found the result had the wrong sign reducing interest in the bag model. These results all appear in Physical Review papers. In the 2000's he repeatedly pointed out the shortcomings of SuperString theory and showed that The Standard Model's form could be derived from space-time geometry by an extension of Lorentz transformations to faster than light transformations. This deeper space-time basis greatly increases the possibility that it is part of THE fundamental theory. Recently, Blaha showed that the Weak interactions differed significantly from the Strong, electromagnetic and gravitation interactions in important respects while these interactions had similar features, and suggested that ElectroWeak theory, which is essentially a glued union of the Weak interactions and Electromagnetism, possibly modulo unknown Higgs particle features, be replaced by a unified theory of the other interactions combined with a stand-alone Weak interaction theory. Blaha also showed that, if Charmonium calculations are taken seriously, the Strong interaction coupling constant is only a factor of five larger than the electromagnetic coupling constant, and thus Strong interaction perturbation theory would make sense and yield physically meaningful results.

In graduate school (1965-71) he wrote substantial papers in elementary particles and group theory: The Inelastic E- P Structure Functions in a Gluon Model. Phys. Lett. B40:501-502,1972; Deep-Inelastic E-P Structure Functions In A Ladder Model With Spin 1/2 Nucleons, Phys.Rev. D3:510-523,1971; Continuum Contributions To The Pion Radius, Phys. Rev. 178:2167-2169,1969; Character Analysis of U(N) and SU(N), J. Math. Phys. 10, 2156 (1969); and The Calculation of the Irreducible Characters of the Symmetric Group in Terms of the Compound Characters, (Published as Blaha's Lemma in D. E. Knuth's book: *The Art of Computer Programming Vols. 1 – 4*).

In the early 1980's Blaha was also a pioneer in the development of UNIX for financial, scientific and Internet applications: benchmarked UNIX versions showing that block size was critical for UNIX performance, developing financial modeling software, starting database benchmarking comparison studies, developing Internet-like UNIX networking (1982) and developing a hybrid shell programming technique (1982) that was a precursor to the PERL programming language. He was also the manager of the AT&T ten-year future products development database. His work helped lead to commercial UNIX on computers such as Sun Micros, IBM AIX minis, and Apple computers.

In the 1980's he pioneered the development of PC Desktop Publishing on laser printers. and was nominated for three "Awards for Technical Excellence" in 1987 by PC Magazine for PC software products that he designed and developed.

Recently he has developed a theory of a Megaverse – containing universes of which our universe is one – with quantum particle-like properties based on the Wheeler-DeWitt equation of Quantum Gravity. He has developed a theory of a baryonic force, which had been conjectured many years ago, and estimated the strength of the force based on discrepancies in measurements of the gravitational constant G. This force, operative in D-dimensional space, can be used to escape from our universe in "uniships" which are the equivalent of the faster-than-light starships proposed in the author's earlier books. Thus travel to other universes, as well as to other stars is possible.

Blaha also considered the complexified Wheeler-DeWitt equation and showed that its limitation to real-valued coordinates and metrics generated a Cosmological Constant in the Einstein equations.

The author has also recently written a series of books on the serious problems of the United States and their solution as well as a book on the decline of Mankind that will follow from current social and genetic trends in Mankind.

In the past twelve years Dr. Blaha has written over 40 books on a wide range of topics. Some recent major works are: *From Asynchronous Logic to The Standard Model to Superflight to the Stars*, *All the Universe!*, *SuperCivilizations: Civilizations as Superorganisms*, *America's Future: an Islamic Surge, ISIS, al Qaeda, World Epidemics, Ukraine, Russia-China Pact, US Leadership Crisis*, *The Rises and Falls of Man – Destiny – 3000 AD: New Support for a Superorganism MACRO-THEORY of CIVILIZATIONS From CURRENT WORLD TRENDS and NEW Peruvian, Pre-Mayan, Mayan, Anatolian, and Early Egyptian Data, with a Projection to 3000 AD*, and *Mankind in Decline: Genetic Disasters, Human-Animal Hybrids, Overpopulation, Pollution, Global Warming, Food and Water Shortages, Desertification, Poverty, Rising Violence, Genocide, Epidemics, Wars, Leadership Failure.*

He has taught approximately 4,000 students in undergraduate, graduate, and postgraduate corporate education courses primarily in major universities, and large companies and government agencies.

www.ingramcontent.com/pod-product-compliance
Lightning Source LLC
Chambersburg PA
CBHW082009190326
41458CB00010B/3128